Guide to Self Patenting

Guide to Self Patenting

Albert Thumann, P.E.
Richard C. Hsu, Patent Attorney

Published by
THE FAIRMONT PRESS, INC.
700 Indian Trail
Lilburn, GA 30247

Library of Congress Cataloging-in-Publication Data

Thumann, Albert.
Guide to self patenting / Albert Thumann, Richard C. Hsu.
p. cm.
Includes index.
ISBN 0-88173-250-8
1. Patent laws and legislation--United States--Popular works. I. Hsu, Richard C. II. Title.
KF3114.6.T48 1997 346.7304'86--dc21 96-29459
CIP

Published by The Fairmont Press, Inc.
700 Indian Trail
Lilburn, GA 30247

Printed in the United States of America

10 9 8 7 6 5 4 3 2 1

0-88173-250-8 FP

0-13-888538-9 PH

Distributed by Prentice Hall PTR
Prentice-Hall, Inc.
A Simon & Schuster Company
Upper Saddle River, NJ 07458

Prentice-Hall International (UK) Limited, London
Prentice-Hall of Australia Pty. Limited, Sydney
Prentice-Hall Canada Inc., Toronto
Prentice-Hall Hispanoamericana, S.A., Mexico
Prentice-Hall of India Private Limited, New Delhi
Prentice-Hall of Japan, Inc., Tokyo
Simon & Schuster Asia Pte. Ltd., Singapore
Editora Prentice-Hall do Brasil, Ltda., Rio de Janeiro

CONTENTS

Foreword

Everyone at one time has a new idea for a product. Some people produce the item out of their garage while others seek to protect their product through patenting. A patent protects one's right to make, use or sell the creation. More than 100,000 patents are issued annually and more than five million patents have been granted in the United States since the first one was issued in 1790 to Samuel Hopkins for a process to make potash.

The inventive process is not discriminatory. Successful inventors have included housewives, blue- and white-collar workers, and engineers and scientists in industry. These inventors are faced with many crucial decisions, such as deciding if the product is "new" and, if so, determining what steps should be taken to protect their product. The inventor can decide to seek patent protection through the services of a patent attorney or lawyer, can decide to self-patent, can decide that patent protection is not required, or can decide to seek development of the product through an invention developer. If it is decided to self-patent, a thorough understanding of patent law is required. Even if it is decided to hire a patent attorney or lawyer, knowledge of patent law is necessary in order to avoid costly errors.

Engineers and scientists employed in industry find that they are often involved in patent-related problems. If they invent a product or process, will it infringe on another company's patents? If their company is sued for patent infringement, how can they help their company? These are just a few of the challenges they face. The intent of this book is to provide the answers to patent law problems. This book does not advocate that everyone with a patentable product should patent without a lawyer. Rather, it is advocated that by understanding the principles of self-patenting the best course of action can be chosen.

Chapter 1

How to Protect A Patentable Idea

Each year more than 100,000 patent applications are filed in the United States alone. When one considers the worldwide patent systems, it probably will be no surprise to find a good idea being developed by more than one person. The inventive process has made countries great, and has made individuals rich and famous.

The *Official Gazette* of the United States Patent and Trademark Office has been published continually for more than 100 years. It contains more than three million case histories of technical accomplishments.

The *Official Gazette* of the United States Patent and Trademark Office is the official journal relating to patents and trademarks, and is now issued each Tuesday in two parts, one describing patents and the other trademarks. It contains a claim and a selected figure of the drawings of each patent granted on that day; notices of patent and trademark suits; indexes of patents and patentees; list of patents available for license or sale; and much general information such as orders, notices, changes in rules, changes in classification, etc. The *Official Gazette* is sold on subscription and by single copies by the Superintendent of Documents, U.S. Government Printing Office, Washington, D.C. 20402.

The illustrations and claims of the patents are arranged in the *Official Gazette* according to the Patent and Trademark Office classification of subject matter, permitting ready reference to patents in any particular field. Street addresses of patentees and a geographical index of residents of inventors are included. Copies of the *Official Gazette* may be found in public libraries of larger cities.

Today, more than ever, the patent process is alive for inventors of new products to meet the energy demands of the future. Patented inventions such as energy-saving devices are creating many opportunities for entrepreneurs and companies.

The patent system has worked enormously well since it was started. Many inventions in the past were the result of individual genius. Today the emphasis is upon a team effort of engineers, scientists and researchers. The question many people are asking today is whether the protection of an invention by the patent system is in conflict with free enterprise and federal anti-trust laws. A patent gives an inventor a limited monopoly to commercially develop the idea. The personal economic benefit offered by this limited monopoly often stimulates inventions which otherwise might likely never have been developed. The real question is: If the patent system were abolished would companies spend millions of dollars a year for research and development, or would entrepreneurs take the risk to develop their ideas?

THE MARK OF MAN

Dating back to the 13th century, man had a desire to leave his mark, or personal identity, on the goods which he produced. These marks indicated many things, including workmanship, origin of product, ownership and craftsmanship. Although the origin of marks is not known it is thought that they originated in Europe by the Germanic peoples. Many of these marks were carried on by a family for generations. It is rare to find two marks which are exactly alike and there is reason to believe that there could have been a register of marks.

THE CREATION OF THE PATENT

In 1684 the legislature of Connecticut stated: "There shall be no monopolies granted or allowed among us, but of such new inventions as are profitable to the country." This law encouraged new inventions. With the vast natural resources available in the United States, support of the church, the early iron industry, factories and government support, the creation of new inventions and promotion of science abounded.

THE FIRST PATENT LAW

The first Congress passed the first federal patent law in 1790, which was signed by President Washington. Thomas Jefferson became administrator of the first patent act. The first patent was granted to Samuel Hopkins for "a new method of making pot and pearl ashes." When the demand for patents became so great that the Patent Committee could not handle all the petitions, the first patent law had to be revised. In 1793 a second bill was passed which provided that patents could be granted upon the oath of the applicant that his device or invention was original and did not infringe upon the existing patents. The aim of the act was to simplify proceedings, but because it relied only upon the oath of the applicant, many patent suits occurred. The act remained until 1836.

The Patent Act of 1836 provided for compulsory examination of all patent applications. A staff of examiners was established and a library was built to house works of science and scientific periodicals along with a special Patent Office Building. A Register of Patents was also inaugurated. The patent system of 1836 came at a time when the young nation had begun to develop its natural resources and has continued to this day in principle. The first patent granted under this act was for a steam locomotive engine.

Patent records in the pre-Civil War decade indicate America's achievements in industrial, scientific and technological developments. In the year of 1850, 2,183 patents were issued. Most of the patents involved labor-saving machinery. During this period a mania for inventing swept the country. Abraham Lincoln even got into the act. He was granted a patent in 1848 for a device for buoying vessels over shoals and sandbars.

By 1850, the new Patent Office that was built in 1836 was overcrowded, partly due to the fact that patent applications were accompanied by models. By 1850 there were more than 7,000 models which had accumulated during the preceding 14 years.

ENTREPRENEURS OF THE PAST

Ole Evinride was the son of a Norwegian immigrant farmer. Very few people would recognize his name today, but he was responsible for designing and building a two-cycle, single-cylinder, air-cooled, gasoline-powered contraption that could be hitched onto the rear of a rowboat. Ole married his girlfriend Bess and founded a company. Their initial slogan was "Throw Away Your Oars! Buy an Evinride Detachable Rowboat Motor." This company grew into the Outboard Marine Corporation, which last year had sales of nearly $500 million.

Another name which nearly everyone will not recognize is A.M. Bute. Bute grew tired of getting up in the middle of the night to put more coal in the furnace. In 1883 he invented a damper flapper to prevent this by changing the heat output of the furnace through a motorized damper and thermostat. The result of his efforts launched a company which, after several mergers, represents the multimillion-dollar Honeywell Corporation.

WHAT IS A PATENT?

A patent grants its owner the right to exclude others from making, using or selling the claimed invention in the United States without a license. A patent does not necessarily grant the owner the right to practice the invention, since this might require the use of technology patented by someone else. Patents are considered the strongest form of intellectual property since the original creation is not a defense to patent infringement. Patents are also time-consuming and expensive to obtain, leading many to only seek them for important and profitable inventions.

Patent protection, prior to June 8, 1995, lasted for 17 years from the day the patent was issued. For patent applications filed on or after June 8, 1995, the term of a U.S. patent begins on the date the application (or any earlier-filed related applications) was filed. Under transitional provisions, patents that issue from an application that was pending prior to June 8, 1995, and patents whose terms have not expired by June 8, 1995, are entitled to either the "17 years from issue date" term or the "20-years-from-earliest filing date" term, whichever is longer.

WHAT IS PATENTABLE?

Under U.S. laws, a patent can be issued to cover any "new and useful process, machine, manufacture, or composition of matter, or any new and useful improvement thereof." 35 U.S.C. § 101. Basically, one can patent any tangible subject matter or, in words of the United States Supreme Court, "anything under the sun that is made by man." Examples of what are not patentable include laws of nature, physical phenomena and abstract ideas. Similarly, functions cannot be patented. Only the *means* or *structure* that performs the function or the *composition* of matter or ideas can be patented. The distinction is placed on the means rather than the function. In other words an idea for creating new energy forms is not patentable, but a workable machine which produces that energy form is patentable. The claims of the patent are confined to the mechanisms and processes which produce the end result.

A patentable invention need not be a technological milestone, like the Shockley transistor or the Hughes rock drill bit. One can obtain a patent on even a modest improvement in the existing technology as long as the improvement would not be obvious to someone who is skilled in the relevant technology.

REQUIREMENTS FOR OBTAINING A PATENT

In order to receive the protection of a patent, an invention must be useful, novel and nonobvious.

Utility

In order for an invention to be patentable, it must be *useful*. In other words, it must be capable of producing a result that is beneficial in some degree to somebody. Useless articles are not patentable. If an invention harms the health of the public, it is not patentable. An invention is considered injurious to the public health only if the injuries cannot be avoided.

The *degree* of utility of an invention and the number of persons it is useful to are immaterial to the patent law.

Except in the cases of chemical inventions, the test is more or less perfunctory, requiring nothing more than the suggestion that the invention might have some practical use. For a product to be useful it should at least work, although it does not have to work perfectly or even better than any competing products or processes that might exist. Moreover, there exists a presumption in favor of usefulness, that is to say that an invention is considered useful unless there is some reason to believe that it will not work, as for example, a perpetual-motion machine.

Novel

The requirement that an invention be novel simply means that the patent applicant must be the first person to invent the subject matter of a patent.

The definition of new as used in this statute is more restrictive than in a dictionary definition. "New" means novel in the Patent Law, not just freshly made or unused. Thus, to be patentable, a thing must have a characteristic which differentiates it from anything that preceded it. A thing can be new in the ordinary sense of the word but not novel in the Patent Law sense. The Patent Law tries to prevent people from copying an existing invention instead of inventing it themselves, no matter how or where the copying takes place. The mere existence of a prior foreign invention is not enough in itself to negate novelty unless the inventor had prior knowledge of the foreign patent.

Determining whether a product or process is within the patent statute's meaning of "novelty" requires comparing the prior art, which is a collection of any past patents and relevant technical literature, with the subject matter sought to be patented. A product is not considered "novel" if all the elements of that product were disclosed in a single piece of prior art.

For an invention to be novel, it also cannot have been invented if the inventor fails to file a patent application within one year after first publicly using, describing or commercializing the invention. Since even so-called "Beta releases" and confidential product proposals can sometimes trigger this one-year "statutory bar," many people lose their U.S. patent rights by failing to file a timely patent application.

Nonobvious

The third, and usually the most critical, hurdle in obtaining a patent is that the invention be nonobvious. The basic rule is that an invention is not patentable "if the differences between the subject matter sought to be patented and the prior art are such that the subject matter as a whole would have been obvious at the time the invention was made to a person having ordinary skill in the art.—35 U.S.C. §103."

In other words, the innovation must be a substantial advance over what was known in the field before; that is to say that if, in light of pre-existing technology, publications and other sources of public knowledge, the invention would have been obvious to one of ordinary skill in the relevant field of the invention, it would not be patentable.

Merely changing the size, proportion, form, arrangement or location would not entitle an invention to be patentable. Neither would duplication or reversal of parts, making devices portable or reversal of movements. But if these changes bring about a new or unexpected and useful result or a new mode of operation, then the invention may be patentable. With the current appreciation for costly energy, many inventors have tried to patent energy-efficient devices. To their dismay they find that their product was invented 20 or more years ago.

In 1966, the Supreme Court, in *Graham v. John Deere Co.*, 383 U.S. 1 (1966), interpreted and applied section 103. The Supreme Court stated:

> Under 103, the scope and content of the prior art are to be determined; differences between the prior and the claims at issue are to be ascertained; and the level of ordinary skill in the pertinent art resolved. Against this background, the obviousness or nonobviousness of the subject matter is determined. Such secondary considerations as commercial success, long-felt but unsolved needs, failure of others, etc., might be utilized to give light to the circumstances surrounding the origin of the subject matter sought to be patented.

Thus, a conclusion of obviousness may be refuted by showing secondary considerations such as unexpected results or benefit, commercial success, long-felt need or the failure of others to achieve the invention.

A SIMPLE, LEGALLY EFFECTIVE WAY OF PROTECTING THE INVENTION

The best way to protect a patentable idea is to keep a good patent notebook. In many patent-related litigations it is a matter of proving who invented the subject matter first in order to decide who should be granted the patent. A slipshod patent notebook can contribute to a loss of royalty income, loss of patent rights and expenditure of heavy legal fees. A good patent notebook discourages litigation by indicating to another party that the inventor has unshakable proof of prior dates of conception and reduction to practice.

A patent notebook may be hard-covered, permanently bound with numbered, stitched pages. There must be no doubt in the court's mind that pages were neither added nor deleted.

There should be no erasures that throw doubt on the integrity of the entire notebook, and pages should not be removed. In the case of corrections, matter which is wrong should be visibly stricken out and corrections made, initialed and dated by the inventor. These notebooks should also be witnessed.

The language used in the notebook must be clear and all key dates must be indicated. Critical dates include:

- When the invention was conceived;

- When the invention was first disclosed to witnesses, together with their signatures;

- When the invention was reduced to practice to prove when the device was constructed (if never reduced to practice, the filing date of the application is important);

- Significant events in the development of the invention.

The date of conception is the time when the inventor thought of a definite method of solving a problem. At this time the inventor should make a clear drawing of the invention and show how it works. Also, a full description of the invention, explaining how it works and why it will be an improvement on existing devices, should be made. The drawings and descriptions should be neat, clear and thorough because they may have to be used in court as evidence to prove the time of conception. Among the items to be included are:

1. What was wrong in the prior art.

2. How this can be corrected by the present invention.

3. The details of the present invention and how it works, including all illustrations for additional advantages and disadvantages of the new invention.

4. A description of operation, and a list of what is believed to be new.

Drawings are also a valuable part of proving conception. The title box of a drawing should be identified by the draftsman who worked on it. The date should appear and it should be possible from the drawing to determine when the drawing was started and when the drawing was completed.

A good collaborating witness is one who is thoroughly capable of understanding the invention. The person should have taken the time to review the invention and sign at the end of the disclosure with a statement such as: Name of Witness, Date. In determining which of several inventions was conceived first the knowledgeable witness will be thoroughly familiar with the subject matter and will be able to answer questions about it. The witness should not be a co-inventor or have any financial interest in the patent sought. It is suggested that two witnesses sign the disclosure in order to increase the probability of finding one if you need this individual years later. It is also recommended that the witness sign and date all relevant notebook papers so that they can be introduced as evidence.

The results of a slipshod notebook can best be seen in the landmark case in the 1960s when Gordon Gould challenged the validity of the basic laser patent 2,929,922. The court ruled that Gordon's notebook was too ambiguous to justify the conclusion that he possessed a definite and permanent idea of the complete and operative invention or that he made his invention sufficiently plain to enable those skilled in the art to understand it. Gould's notebook did not clearly indicate what activities were performed and their dates, and how they pertained to reducing the invention to practice. Finally, in 1977, after many appeals and providing additional data, this ruling was overturned. Almost 18 years after Gould invented the laser amplifier, he was granted a patent for it. Gould is expected to receive $200 million in royalties over the 17-year life of the patent. The importance of considering the notebook a legal entity and providing sufficient documentation could save years of costly litigation and a large sum in royalties.

Figure 1-1 summarizes the guidelines for a legally effective notebook. Patent notebooks are available from most office supply stores. Several patent notebooks will be required to illustrate the details properly. Thomas Edison used 3,400 notebooks during his career.

USE OF AGREEMENTS

In many instances the inventor needs other people to produce drawings, prototypes, etc., during the conceptual stage of an invention. The inventor should take steps to insure that the idea will not be taken by others. One method of protection is to use a simple agreement form made up between the inventor and the second party. The agreement states that the invention has been disclosed to the other party for the sole purpose of making a drawing, a prototype, etc. A typical agreement between a draftsman and the inventor is illustrated in Figure 1-2.

DISCLOSING THE INVENTION FOR MANUFACTURING FEASIBILITY

Many times the initial invention as specified is not practical to manufacture in its present form. It is best to find this out *prior* to applying for a patent. Remember the patent is granted based on the specified structure which is designated by the claims.

In order to obtain manufacturing feasibility it may be necessary to build a prototype. Any developments which result during

- Use a numbered, bound notebook.
- Avoid blank spaces. Draw an "X" through blank spaces. Sign and date each page.
- Don't change a page after it is signed and dated. To update information use a new page and reference the original page.
- Don't erase or remove pages from notebook. Draw lines through unwanted items so as not to destroy legibility.
- All entries should be made in black ink.
- All entries should be in chronological order.
- Include all references to any sources of information.
- Sign and date all photos or drawings and reduce them so they can be inserted into the book. The signature should overlap the affixed drawings on the page to prove it was not added at a later date.
- All ambiguities, code names, etc., should be identified and defined.
- Have one or more witnesses who understand your invention sign and date time of disclosure on every page after they have read and understood contents.

Figure 1-1. Guidelines for a legally effective notebook.

I. . . . .
Name of Draftsman

hereby certify that on . . . . .19 . . . . . . I was employed to produce drawings of an invention relating to

. . . . for . . . .
Invention Name of Inventor

from sketches and drawings and description furnished by the inventor and that I fully understand the construction and operation of this invention, that I disclaim being the inventor or co-inventor of this invention, and that I completed each drawing on the date written by me for that drawing.

. . . .
Signature of Draftsman

. . . . Date

Figure 1-2. Typical agreement between a draftsman and inventor.

the prototype phase can then be incorporated into the design of the patent. The inventor may require the expertise of someone familiar with building prototypes or a manufacturer of a similar product. Again the inventor should protect his invention by having an agreement signed. A typical secrecy agreement is illustrated in Figure 1-3.

The inventor of a . . . . discloses product to the undersigned for the sole purpose of obtaining manufacturing feasibility. The product will be held in the strictest confidence and will not be used for any other purpose by the undersigned, the company he represents or any party assigned by him. Further it is agreed that the undersigned disclaims being the inventor or co-inventor of this invention and any developments or improvements are the property of the inventor.

. . . .
Inventor

. . . .
. . . .
company represented

. . . .
date

Figure 1-3. Agreement of secrecy.

DISCLOSING IDEAS TO LARGE COMPANIES

Most companies have established policies which protect their proprietary research. When an outside inventor submits a product for possible licensing, many companies wish to establish that in fact they may already be working on this idea and are not liable in any way for the disclosure made by the inventor. Figure 1-4 illustrates a typical form letter which must be signed by the outside inventor prior to disclosing the invention. This type of agreement discourages invention disclosures unless the inventor has filed a patent application. It is very difficult for the inventor to have a large company sign a disclosure agreement which the inventor

Date: ____________

_______________ CORPORATION
CONDITIONS FOR SUBMISSION OF TECHNICAL IDEAS

__________ shall not be obligated to treat as confidential any information you may disclose to us, and you will rely for protection of your interests solely upon your rights under the Patent Laws; it being understood that __________ shall have no obligation or liability of any kind except as the same may arise under such laws.

Disclosure of your idea to __________ shall be made in writing accompanied by appropriate drawings or sketches—all signed, dated and witnessed. Preferably disclosure should be made by sending us a copy of any patent application you may have filed covering your idea, or any issued patent covering it.

Disclosure documents furnished to __________ need not be returned to you, so that we may have a record of your disclosure; and we may make copies of the same should we consider this advisable in evaluating your idea.

Such consideration as we give your idea shall be controlled solely by our judgment; we shall not be obliged to give reasons for such decisions as we may make concerning it; and under no circumstances shall we have any obligation to reveal to you any information regarding our Company's activities.

No obligation or liability of any kind shall be assumed by or implied against __________ or any of its officers or employees unless and until a formal written agreement is reached between you and __________ and then the sole obligation will be that imposed by such agreement.

The undersigned represents and warrants that he has the right to make such disclosure and that he is, with respect to the subject matter thereof, the only party in interest.

I hereby confirm my agreement to the foregoing:

Signed ____________________ Date______________

Witnessed ________________

Figure 1-4. Form letter for disclosing ideas to a large company.

has drawn up to protect the idea. Thus, if an inventor wishes to license the invention to a large company it is extremely important to have filed a patent application first in order to properly protect the idea.

SUMMARY OF PATENTABILITY

Whoever invents any new and useful invention may get a patent on it. The requisites of patentability are utility, novelty and nonobviousness. The court's interpretation of these three words has played an important role in the development of patent law.

Chapter 2

What to Know About Choosing a Patent Lawyer, Agent or Invention Developer

Patent law requires that an application be prosecuted either by the inventor or by a registered patent attorney or agent. Registered means that the attorney or agent has performed the office requirements and has been officially admitted to practice before the Patent Office.

WHAT IS THE DIFFERENCE BETWEEN PATENT ATTORNEYS AND PATENT AGENTS?

A registered patent agent need not be a lawyer. A better name for patent agent would be "patent solicitor." A patent agent cannot appeal a U.S. Examiner's decision in the U.S. Court of Customs and Patent Appeals, nor file an infringement suit, nor draw up contracts such as assignments of licenses relating to patents. In other words, work is confined only to the Patent Office. A patent lawyer or attorney, on the other hand, is educated in law, has passed the bar examination and has complied with other requirements for being licensed as a lawyer in that jurisdiction, in addition to passing the Patent Office requirements. Thus, a patent lawyer can represent an inventor in appealing an examiner's decision, fighting an infringement case, or drawing up contracts. The only similarity is that patent agents or patent lawyers are both registered by the Patent Office. Before 1938, persons registered to practice before the Patent Office were listed as attorneys. After that date, persons who were not lawyers, members of the bar, were listed as patent agents.

The Patent and Trademark Office cannot recommend any particular attorney or agent, or aid in the selection of an attorney or agent, as by stating, in response to inquiry, that a named patent attorney, agent or firm is "reliable" or "capable." The Patent and Trademark Office publishes a directory of all registered patent attorneys and agents who have indicated their availability to accept new clients, arranged by states, cities and foreign countries. The directory may be purchased in paper from the Government Printing Office or from the PTO's Office of Electronic Information on the CASSIS ASIST CD-ROM disc.

BEWARE OF UNREGISTERED AGENTS

The purpose of this section is not to imply that all unregistered agents or invention developers are frauds, but rather to illustrate some of the cautions the amateur inventor must take to avoid some costly mistakes.

An aspect which may be confusing to the patent applicant is whether or not the person the applicant has contracted is registered by the Patent Office. Until recently one registered to represent an applicant on an official basis at the Patent Office was not allowed to advertise. Usually advertisements in newspapers, radio and television are for organizations which are really "Invention Developers" and are not registered patent agents. Much has been said about invention developers and many of these statements have not been very flattering.

It has been estimated that invention developers make in excess of $100 million from the fees they charge inventors. The real question is whether or not the inventor is being asked to go full speed ahead with the patent application even if it is unwise. Some patent invention developers advertise that for a fee of $2,000 or less they will prepare a patent application and guarantee that the inventor is granted Patent Pending status. The interesting thing is that for a mere $245 anyone can prepare his own patent application and receive Patent Pending status from the U.S. Patent Office. As a test case, three inventors went to an invention development firm and disclosed their invention to the firm. Within seven days they received a letter from the vice president of the firm who indicated that he had completed the preliminary review of the details of the invention and believed the inventors should go ahead with their invention.

The invention developers wanted a fee plus 20 percent commission on net proceeds received as a result of their efforts. The fee included examination of the proposal by their technical and marketing departments with respect to commercial application; preparation of a prospectus by creative writers indicating details that would be of primary importance and interest to prospective manufacturers; and selection of prospective participants with related manufacturing, distributing and licensing interest by the developer's research staff. Copies of the promotional material would be distributed to all prospects and a campaign would be established to publicize the new invention.

All negotiations for sale or licensing of the invention would be handled with interested companies. The pledge of the invention developer usually is to utilize its high-powered program. The invention developer usually emphasizes that the inventor has to deal with small businessmen, various specialists, engineers, lawyers, draftsmen, patent attorneys, marketing organizations and sales agencies. Contracting with the invention developer provides one central contact for dealing with all of these people.

A follow-up letter from the invention developer suggesting financial difficulties often causes the inventor to temporarily shelve an important product, thus risking a possible loss of revenue. A plan is offered whereby the inventor can pay only half the fee at this time and the balance within a certain period of time. Many inventors are pressured into proceeding in this way.

Unregistered agents or developers usually stay within the boundary of the law. They do not say that they will prosecute an application; they only indicate that they will have it prepared. Remember, if the inventor decides to have a patent prepared by a developer, the inventor must sign the application indicating that the inventor is prosecuting the patent. The specification must be adequately understood so that the inventor (not the developer) will be able to defend it later on, directly or indirectly, with the Patent Office.

Inventors usually find Patent Law difficult at first because the subject matter involves ideas rather than material things like land and money. Therefore contracts involving inventions or patents must have a clear understanding of the idea of what is being made the subject of the agreement.

Cases have been found where unethical agents have prepared applications in one day. They may write claims that are too broad or too narrow. The very broad claims are sure to be rejected while the claims which are too narrow are likely to be allowed, but have no commercial value. At a later time, when the application comes back with the comments from the Patent Examiner without further prosecution, these "patent factories" cancel the rejected claims and send in the final fees. The inventor gets nothing but an impressive-looking certificate and a bad experience.

From this discussion it is apparent that unless the inventor is dealing with a reputable patent lawyer or attorney a substantial risk is taken. Therefore the inventor must choose carefully. If he or she feels strongly that the invention is very marketable and too complex for self-patenting, then the utmost protection should be employed and a good patent attorney should be used. If, on the other hand, the invention is simple and the inventor has a good grasp of the principles of patent law involved, he may decide to take the risk and file a patent application without a lawyer.

IS IT ADVISABLE TO TRY TO SELL AN INVENTION WHERE NO APPLICATION HAS BEEN FILED?

Invention developers many times advertise schemes for selling inventions where no patent applications have been prosecuted. The invention is said to be protected by the records of the inventor or by disclosure statements signed by the prospective licensing company. Records alone do not give exclusive rights to the invention. The records only fortify the inventor's position in case the patent application is thrown into interference proceedings. As indicated in Chapter 1, most large manufacturers do not sign agreements which waive their rights to use an idea before they know what it is. It might turn out that the idea is old and unpatentable or it might be an idea that their own engineers are working on. In either case manufacturers do not want to risk giving up something they have a perfect right to. The inventor should not describe an invention where no application has been filed. The inventor should proceed with the patent application to get Patent Pending status and then try to sell the invention to a manufacturer.

CHOOSING A PATENT LAWYER

All things being equal, it is best to obtain a patent lawyer who has an office near the inventor. The Code of Federal Regulations, Title 37, § 1.2 states:

> **§ 1.2 Business to be transacted in writing.**
> All business with the Patent and Trademark Office should be transacted in writing. The personal attendance of applicants or their attorneys or agents at the Patent and Trademark Office is unnecessary. The action of the Patent and Trademark Office will be based exclusively on the written record in the Office. No attention

will be paid to any alleged oral promise, stipulation, or understanding in relation to which there is disagreement or doubt.

The inventor can write to the American Patent Lawyer Association or the U.S. Patent Office for names and addresses of members in his or her state. In choosing a patent attorney, ethical standards are important. If a patent attorney doubts that an invention is patentable or points out that the market is small, or that the margin of profit is slim, do not conclude that the attorney is a fool and rush out to find another one who will guarantee to get a patent. Remember that a patent attorney makes money by applying for a patent whether or not it is an unpatentable invention or a worthless patent. Therefore, if the attorney says that the invention is not good to patent, the attorney probably has high standards.

The U.S. Patent Office offers many services on the Internet. The home page for the U.S. Patent Office is www.uspto.gov. This home page includes an index of 18,000 attorneys and agents licensed to practice before the U.S. Patent and Trademark Office. A search for attorneys in your area can be made and the entire list can be downloaded.

Many times an inventor who is discouraged to patent by an attorney goes to one who is less particular. An ethical attorney will refuse an application if it is believed that a patent will probably not be granted or the attorney is prosecuting a similar invention for somebody else or the product is not marketable. The ethical patent attorney wants a lot of information about the invention before deciding to take the case. The inventor who is knowledgeable in patent procedures can best evaluate the abilities of the patent lawyer.

HOW MUCH WILL IT COST?

Most patent lawyers divide their fee into several categories, namely: search, drawings, fee for application, first amendment and second amendment. Thus the final fee depends on the individual invention and how long it takes until the final patent is granted. Additional fees which sometimes occur are for interferences which may arise or patent infringement cases. It is likely that for a relatively simple invention the total fee from the search through second amendment could be as much as $2,000.

WHAT IS THE POSSIBILITY THAT A PATENT ATTORNEY MIGHT WANT TO STEAL PART OF YOUR INVENTION?

It is very unlikely that a patent attorney will try to steal part of the invention by claiming that some contributions were made to it and therefore he is entitled to be co-inventor and share in the proceeds. It is advised that any disclosures given to the inventor by a registered attorney or lawyer are considered property of the inventor. In practice many patent lawyers make contributions to an invention.

DOES THE SMALL INVENTOR HAVE A CHANCE?

Approximately 75 percent of the patents granted went to corporations with research staffs. The average cost of filing a patent complete with legal and patent application fees is approximately $2,000 to $8,000 nationwide. Complex patent applications and fees can cost $10,000. It is no wonder many small inventors think twice before filing. An additional factor is that if the patent is infringed upon the only recourse for the small inventor is to sue. Patent litigation cases have been known to cost in excess of $50,000. Therefore the question arises, What does the inventor plan to do with the patent? If the inventor wishes to file a patent application for ego purposes, certainly a lawyer is not required. If the inventor plans to license the patent, based on Patent Pending Status, it may or may not be worthwhile to secure an attorney, depending on the complexity of the invention and the inventor's knowledge of patent law. The small inventor has a better chance for success if he understands how patent law works, and is convinced of the novelty of his invention and how it will be marketed.

Chapter 3

The Search

THE IMPORTANCE OF THE PAST

It is rare to find an invention which has no resemblance to already-patented inventions. Of all the patents granted, only a small fraction actually became marketable products. By searching past patents it is possible to spark creative new ideas, use patents where the protection term has expired, or find out if the current idea is really new.

IDEAS FOR NEW INVENTIONS

Many times a would-be inventor gains insight into a new product by merely reviewing past patents. The review of old patents can trigger insight or reduce an individual's labors towards solution of a problem having no relationship with the existing patent. Lawyers have always used published case histories for guidance in the conduct of current cases. These case histories give precedents showing how similar matters were handled. Engineers have a similar reference at their disposal. In more than 100 years, more than three million case histories for technical accomplishments have been recorded. In order to create, it is important to recognize that intelligence is not always the key factor. It is far more important to exercise a childlike imagination and to have an open, unfettered curiosity. An inventor can be thought of as an engineer who doesn't take his education too seriously. Random browsing through a wide variety of patents or literature can stimulate the mind and excite creative imagination.

THE IMPORTANCE OF THE SEARCH

A search is not obligatory, but many times a search will determine whether or not to spend additional money to file a patent application. A search is a good idea unless the inventor is thoroughly familiar with what's already been patented. A patent search through an attorney may cost at least $750. With the average time to get a patent awarded from date of filing 19 months the search can save both time and money.

HOW TO CONDUCT YOUR OWN SEARCH

Today there are many cost-effective ways to conduct your own search. The Scientific and Technical Information Center of the Patent and Trademark Office has a search room open Monday through Friday where the public can examine U.S. Patents granted since 1836. The office is located at Crystal Plaza 3, 2021 Jefferson Davis Highway, Arlington, Virginia. Patents are arranged in 400 classes and 120,000 subclasses. A file information room allows inspection of issued patent files and other open records. This site also includes 120,000 volumes of scientific and technical books in various languages and official journals of 77 foreign patent organizations representing 12 million international patents. In addition major U.S. libraries are designated as patent depository libraries where preliminary searches can be made.

USE OF INTERNET

The Internet is a valuable source of all information concerning patenting. Figure 3-1 indicates 146,203 listings under the search word; patent search.

The U.S. Patent and Trademark Office has a home page on the Internet: www.uspto.gov. Complete information on how to patent and data bases are available through the home page. Typical questions and answers about the data base and how to search are illustrated in Figure 3-2.

Documents 1-25 of 146,203 matching "patent search."
Show summaries for these documents.

- Internet patent search/copy service online, design/...
- STO's Internet Patent Search System
- Boris Pevzner - Home page at MGM
- discount patent search
- Michael J. Colitz, Jr., Patent Attorney - Patentability...
- cd-rom
- So You Have An Idea
- Answers to Frequently Asked Questions About Patents and...
- FAQs: Index
- patentability search necessary
- Inventor's Helper Letter
- Avoiding Patent, Copyright and Trademark Problems
- Affordable Patent Searching, Affordable Trademark Searching
- http://www. tripod.com/%7ENEWWORLD/index. html
- How to Use the CNIDR Mail-Z39.50 Gateway to Search Patents
- U.S. Patent Advanced Search Page
- U.S. Patent Simple Search Page
- U.S. Patent Advanced Search Page
- U.S. Patent Boolean Search Page
- Patenting Help
- SBA: Hotlist
- Intellectual Property: Patents
- Inventor's Awareness Group
- MacMillan Lytle Fisher's Recommended Sites: Business &...
- Important Links

Figure 3-1. Typical listings on the Internet.

What information is in the database?

The database consists of the front pages of U.S. patents (including design patents). The front page of a patent includes all bibliographic data, such as the inventor's name, the patent's title and the assignee's name, as well as the abstract, a brief description of the patent's contents.

What years does it cover?

Currently, the database contains all the patents from 1976 until February, 1996. There are no plans from the Patent Office to include years before 1976 on our service. For years before 1976, a commercial service is your only recourse.

Here's what the PTO has to say on the matter:

> Patent text data is available starting in August of 1971. However, prior to 1976, data files are incomplete. That is, not all of the patents were captured electronically between 1971 and 1975. For this reason, we decided to limit the Internet data to 1976 forward. Users wanting to do a complete patent search should seek other services, including those offered by the PTO's network of 80 Patent and Trademark Depository Libraries located throughout the country.

How often is the database updated?

We update the database as soon as the Patent Office sends us updates, hopefully about once a month. This means that if you are looking for a patent which was issued only in the past few weeks, we probably won't have it in our database yet. Look at the db status page for details.

How do I get the full text and patent images?

The PTO's offering on the Internet includes bibliographic information (including the abstract) for patents granted since 1976. The USPTO has no plans to expand resources available over the Internet to include full text or images. You can, however, order copies of patents to be sent to you through the mail directly from the Patent Office.

For off-line resources, you may want to look into the Patent and Trademark Depository Libraries.

Full text and images *are* available free over the Internet for AIDS-related patents, at the AIDS Patents Database.

Figure 3-2.
Questions concerning the U.S. Patent Office home page.

QUESTIONS ABOUT SEARCHING

How do I use the Boolean query page?

Help is available in the Boolean Search help document.

How do I use the advanced query page?

There is now a document describing how to use the Advanced Query page.

What is the menu under the heading "Which database to search?"

This is the list of databases you can select from. Each database is named by the last two digits in the year that it contains. You can select "95," for example, to search the patents issued in 1995. For databases which cover a range of years, the first two digits are the beginning year, and the last two are the ending year. For example, the '81-85' db covers patents issued between Jan. 1, 1981 and Dec. 31, 1985.

How do I search more than one year at a time?

There are two ways:

1. Select a five-year block of years to search from the **Select Database** menu. The '76-80' db covers 1976-1980, etc.

2.. Search all the years in our database at once by selecting the '76-96' db from the **Select Database** menu.

Note: All the multiple-year databases are currently virtual. This means that searches will take longer, and retrieving additional hits will take as long as the initial search.

How do I search for a particular patent number?

To search by patent number, go to the Boolean Search Page, select which database you want to search in using the **Select Database** menu, type the patent number in the **Term 1** text entry box, and select 'Patent Number' from the **Term 1** fields menu. Then hit the 'submit' button. Note: If you're not sure which year a patent was issued in, search all the years by selecting the '76-96' db.

You can also search for a patent number in the Advanced Query page, using this syntax:

PN/*number*

Where *number* is the number of the patent for which you wish to search. Non-utility patents must include a prefix: 'D' for design patents, 'PP' for plant patents, 'RE' for reissue patents, 'T' for defensive publications, and 'H' for SIRs.

How do I search for a particular class and subclass?

To search by class and subclass, go to the Boolean Search page, select the year you wish to search in from the menu in the lower right hand corner. Select the appropriate Term text entry box and then type the class and subclass in this form:

class/subclass.subsubclass

Choose 'Class/Subclass' from the matching Fields menu and hit the Submit Request button.

You can also search for class/subclass on the Advanced Query page. There is an example demonstrating this on the Advanced Query Help page.

CASE STUDY 3-1

The below case study is based on the author's experience for a new product he invented and thought should be patentable:

The inventors of a telephone mouthpiece refresher described in Figure 3-3 felt they had a new product. What steps should be taken before applying for a patent?

Analysis

The inventors documented their invention and proceeded with a patent search. To the surprise of the inventors, telephone mouthpiece protectors were patented as far back as 1913. The latest patent was issued in 1966. Figures 3-4, 3-5, 3-6, 3-7 and 3-8 illustrate some of the patents discovered. The question facing the inventors is whether the structure of their mouthpiece would be considered not to infringe on previous patents. The inventors decided to get the advice of a patent lawyer. The patent lawyer analyzed the existing patents and advised that even though the patents were similar, he thought that the construction of their product was possibly different enough to justify applying for a patent, but warned that there was a possibility of rejection by the Patent Examiner. The inventors determined that their invention would probably not be allowed and decided to forgo the expense of applying for the patent. A question could be raised if the inventors should have investigated the patentability for the chemical refresher instead of the structure.

TÉLAN

I NAME OF PRODUCT
The product will hereafter be known as TÉLAN.

II PURPOSE OF TÉLAN
TÉLAN is a telephone mouthpiece refresher. It fits over the mouthpiece of a telephone and gives off a pleasant aroma.

III GENERAL DESCRIPTION OF TÉLAN

a. TÉLAN is made in five parts as shown on the attached diagram. The parts are as follows:
    1. Part 1 is the base. It is made of plastic and has five indentations. These indentations hold the aroma refresher tablet.
    2. Part 2 is a filter.
    3. Part 3 is a thin sponge.
    4. Part 4 is a plastic piece which fits over parts 1, 2 and 3 and secures them to the mouthpiece of the telephone receiver. It is made in different colors.
b. TÉLAN will be made in all standard telephone colors: black, white, green, yellow, red, etc.

IV DESCRIPTION OF AROMA REFRESHER TABLET
The tablet used in TÉLAN remains dry although it absorbs almost twice its weight in liquid. The liquid mixture for this tablet involves the use of isopropyl alcohol and formaldehyde in the following quantities:

Isopropyl 10% to 20%
Formaldehyde .5% to 1%
Balance = water

Figure 3-3. Telephone mouthpiece description.

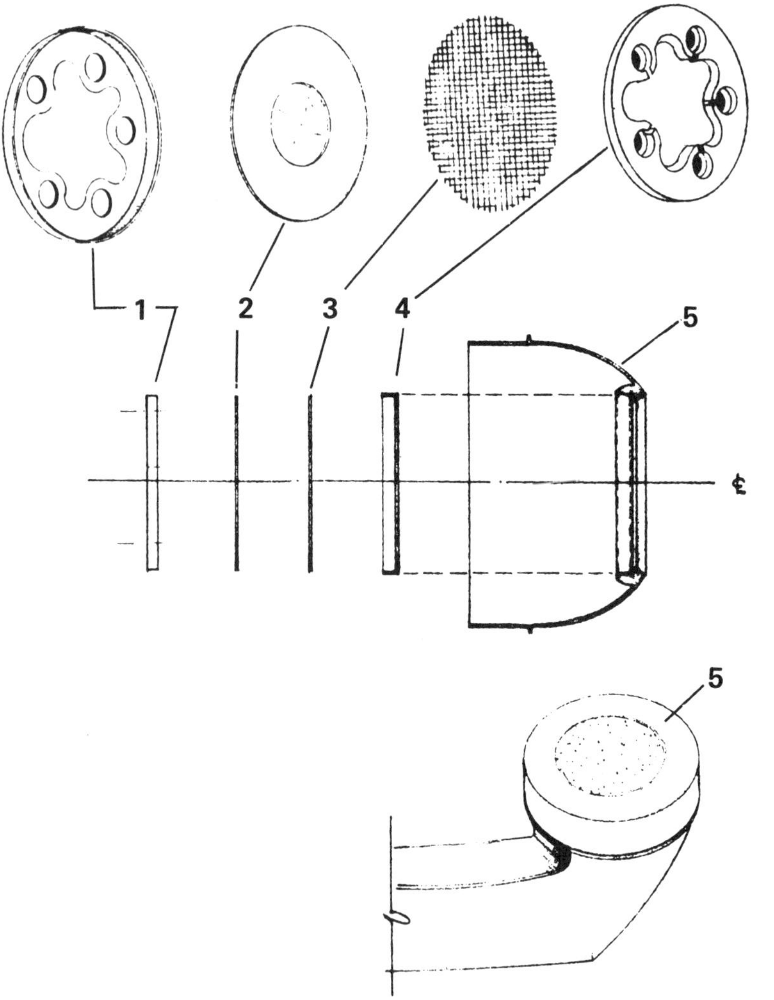

Figure 3-3 (concluded). Telephone mouthpiece description.

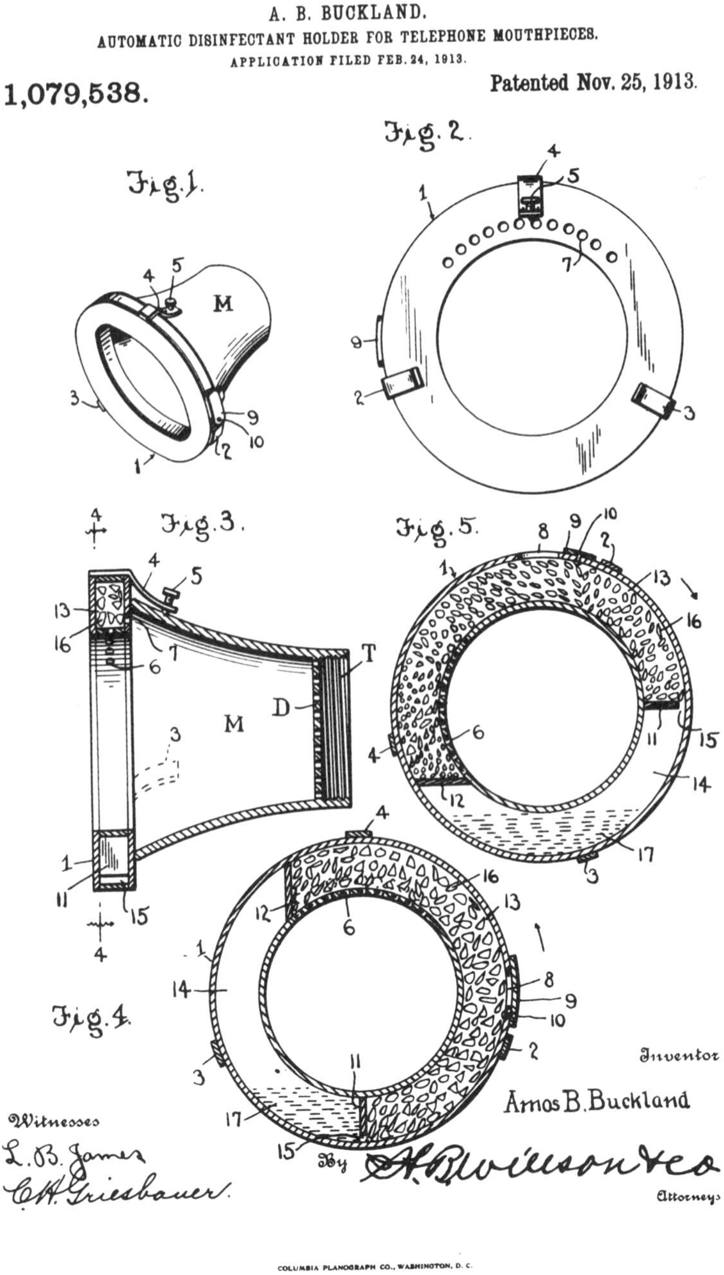

Figure 3-4. Patented telephone mouthpiece.

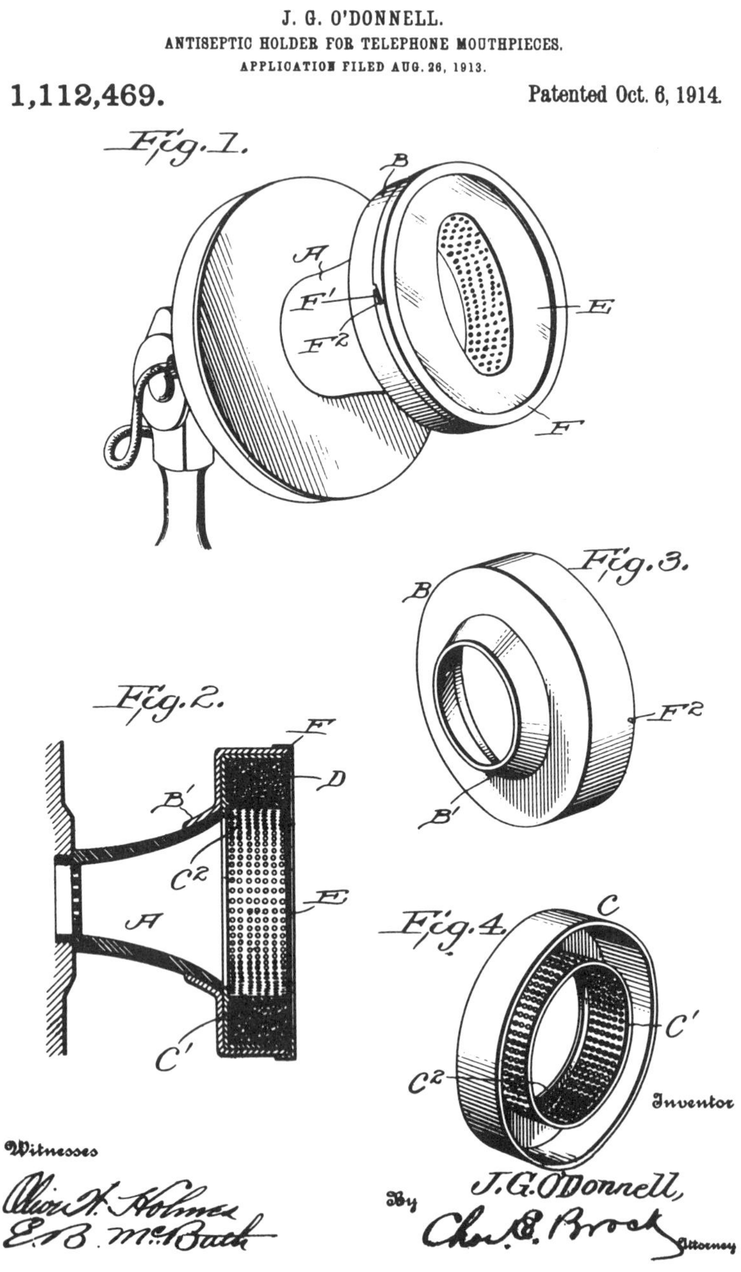

Figure 3-5. Patented telephone mouthpiece.

L. A. COLLINS
TELEPHONE MOUTHPIECE PROTECTOR
APPLICATION FILED MAY 12, 1914

1,140,954. Patented May 25, 1915.

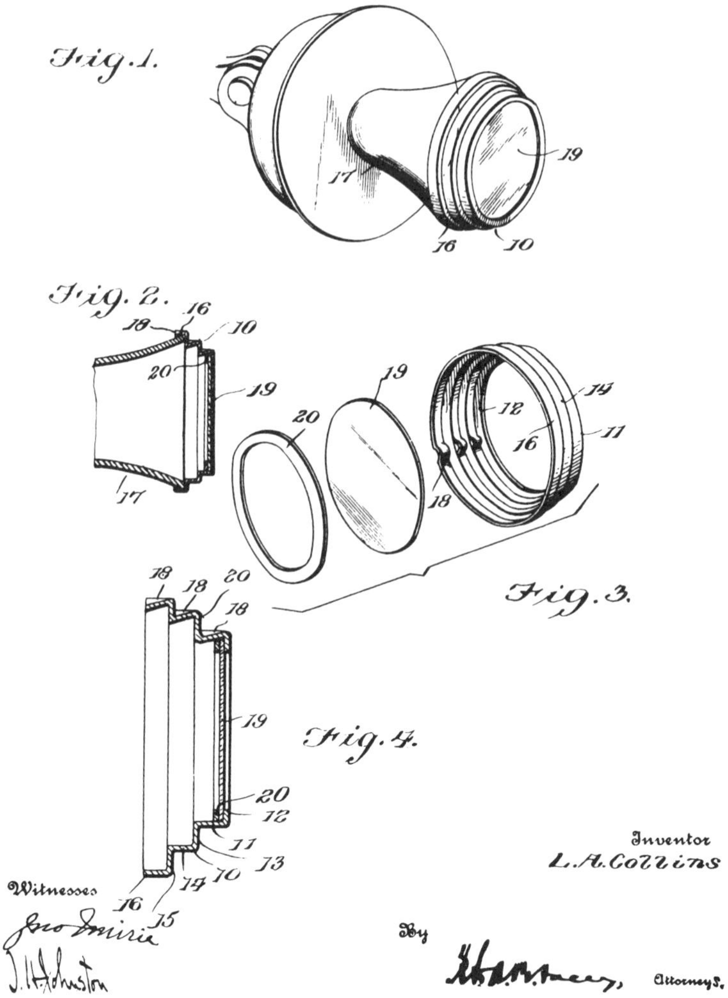

NORRIS PETERS INC. LITHO. WASHINGTON D.C.

Figure 3-6. Patented telephone mouthpiece.

April 15, 1952 C. E. ZIMMERMANN **2,593,382**

SANITARY TELEPHONE MOUTHPIECE ATTACHMENT

Filed June 23, 1950

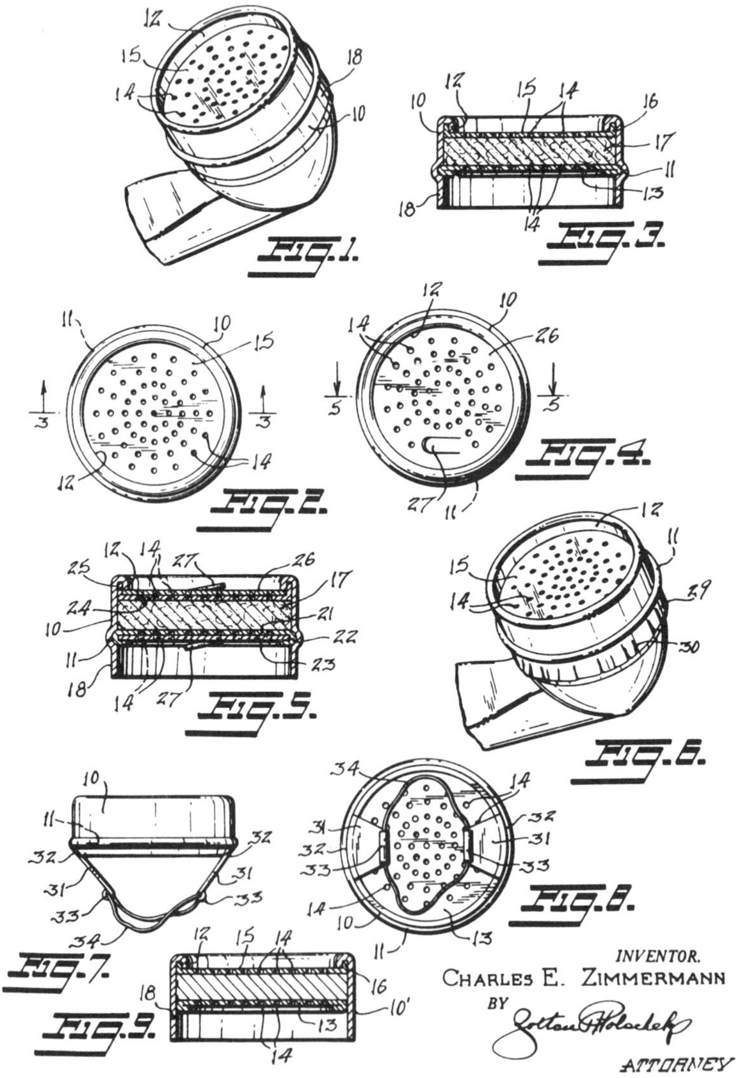

Figure 3-7. Patented telephone mouthpiece.

United States Patent Office

3,238,313
Patented Mar. 1, 1966

1

3,238,313
ANTISEPTIC TELEPHONE MOUTHPIECE GUARD
Theodore P. Kalogris, 2993 Adeline St., Oceanside, N.Y.
Filed Sept. 8, 1961, Ser. No. 136,896
3 Claims. (Cl. 179—185)

This invention generally relates to telephone attachments and has for its primary object to provide novel means for protecting the mouthpiece of the instrument in a manner to prevent the transmittal of various germs.

Another important object of the present invention is to provide, in a manner as hereinafter set forth, an improved antiseptic mouthpiece guard of the character described which is adapted to be readily slipped on a conventional telephone transmitter and removed therefrom when desired.

Still another very important object of the invention is to provide a sanitary mouthpiece guard or attachment of the character set forth comprising a readily removable filter unit of a unique construction whereby the germicidal element may be conveniently replaced from time to time.

Other objects of the invention are to provide an improved antiseptic telephone mouthpiece guard which will be comparatively simple in construction, strong, durable, compact, of light weight, attractive in appearance and which may be manufactured at low cost.

These together with other objects and advantages which will become subsequently apparent reside in the details of construction and operation as more fully hereinafter described and claimed, reference being had to the accompanying drawing forming a part hereof, wherein like numerals refer to like parts throughout, and in which:

FIGURE 1 is a perspective view, showing an antiseptic guard constructed in accordance with the present invention in position on a telephone mouthpiece;

FIGURE 2 is a vertical sectional view, taken substantially on the line 2—2 of FIGURE 1;

FIGURE 3 is a view in front elevation of the filter unit;

FIGURE 4 is a sectional view on an enlarged scale through the filter unit, taken substantially on the line 4—4 of FIGURE 3; and

FIGURE 5 is a perspective view of the device with the component parts thereof separated and showing the holder or cap broken away in section.

Referring now to the drawing in detail, it will be seen that the embodiment of the invention which has been illustrated comprises a generally cup-shaped holder or cap of a suitable elastomeric or other material which is designated generally by reference numeral **6**. The cap **6** includes a skirt **7** which is adapted to be slipped on the usual mouthpiece **8** of a conventional telephone transmitter **9** in a manner to hug and frictionally grip said mouthpiece. This is illustrated to advantage in FIGURE 2 of the drawing.

The holder **6** further includes an end wall **9** with which the skirt **7** is integral. The wall **9** of the cap **6** has formed centrally therein a circular opening **11**. The inner end of the wall of the opening **11** is circumferentially rabbeted as indicated at **12**.

The rabbet **12** of the opening **11** provides a seat for a removable circular filter unit **13**. The filter unit **13** is frictionally retained in the opening **11**. As shown to advantage in FIGURE 4 of the drawing, the filter unit **13** includes separable, telescopic inner and outer rings **14** and **15**, respectively, of a suitable plastic or other material. The outer ring **15** is provided on its outer end with an inturned flange **16**. The inner ring **14** is provided, on its outer end, with an external flange **17** which the ring **15** abuts. Formed integrally with the ring **14** on the inner periphery thereof is a plurality of webs, plates or flanges

2

18. Formed integrally with the plates **18**, on the inner sides thereof, are split rings **19** providing chambers or pockets **20** for the reception of a suitable deodorizing and disinfecting agent. Formed integrally with the flange **16** of the ring **15** are webs or plates **21** which are opposed to the plates **18**. The plates **21** have formed therein circular openings **22** in alignment or registry with the chambers or pockets **20**. The rings **14** and **15** provide a holder for a laminated, germicidal or antiseptic diaphragm or disk **23**. This disk, diaphragm or filter element **23** is suitably chemically treated and, in the embodiment shown, is laminated, comprising an inner disk **24** between relatively large outer disks **25** of a suitable fabric. The peripheral or marginal portion of the disk, diaphragm or filter element **23** is firmly clamped between the ring **14** and the flange **16** of the ring **15**. It also will be noted that the diaphragm or filter element **23** forms a closure for the chambers or pockets **20**.

It is thought that the use of the device will be readily apparent from a consideration of the foregoing. Briefly, with the antiseptic filter unit **13** mounted in the annular seat provided by the circumferential rabbet **12** in the wall of the circular opening **11** in the end wall or flange **10** of the holder or cap **6**, said holder or cap is slipped longitudinally on the telephone mouthpiece **8**. When the device is so mounted the resilient skirt **7** substantially conforms to the mouthpiece **8** and frictionally grips same for retaining the attachment in position thereon. The chemically impregnated germicidal or antiseptic filter element **23**, when the telephone is in use, kills various germs on contact and prevents the spread thereof in an obvious manner. Also, the pockets or chambers **20** defined by the split rings **19** are to contain a suitable deodorizing or disinfecting agent. To renew the filter element from time to time, the cap or holder **6** is removed from the mouthpiece **8** and the unit **13** is then easily removed inwardly from the rabbeted opening **11** in the end wall or flange **10**. The rings **14** and **15** are then separated to permit removal and replacement of the element **23**. Or, if desired, the unit **13** may be discarded and a new unit mounted in the holder or cap **6**. It will be noted that the element **23** closes the openings **22** in the plates **21** of the ring **15** for retaining the agent in the pockets **20**.

The foregoing is considered as illustrative only of the principles of the invention. Further, since numerous modifications and changes will readily occur to those skilled in the art, it is not desired to limit the invention to the exact construction and operation shown and described, and accordingly all suitable modifications and equivalents may be resorted to, falling within the scope of the invention as claimed.

What is claimed as new is as follows:

1. An antiseptic telephone mouthpiece guard comprising a filter unit, said unit including separable, telescopic inner and outer rings, an antiseptic disk removably mounted between said rings, and means for mounting said unit on a telephone mouthpiece, said unit further including plates in one of the rings, and rings on the inner sides of said plates providing pockets for the reception of a deodorizing and disinfecting agent, said antiseptic disk forming a closure for the pockets.

2. An antiseptic telephone mouthpiece guard comprising a filter unit, said unit including separable, telescopic inner and outer rings, an antiseptic disk removably mounted between said rings, and means for mounting said unit on a telephone mouthpiece, said unit further including plates in one of the rings, and rings on the inner sides of said plates providing pockets for the reception of a deodorizing and disinfecting agent, said unit still further including plates on the other ring opposed to the first-named plates and having openings therein aligned with the

Figure 3-8. Patented telephone mouthpiece.

March 1, 1966 T. P. KALOGRIS **3,238,313**

ANTISEPTIC TELEPHONE MOUTHPIECE GUARD

Filed Sept. 8, 1961

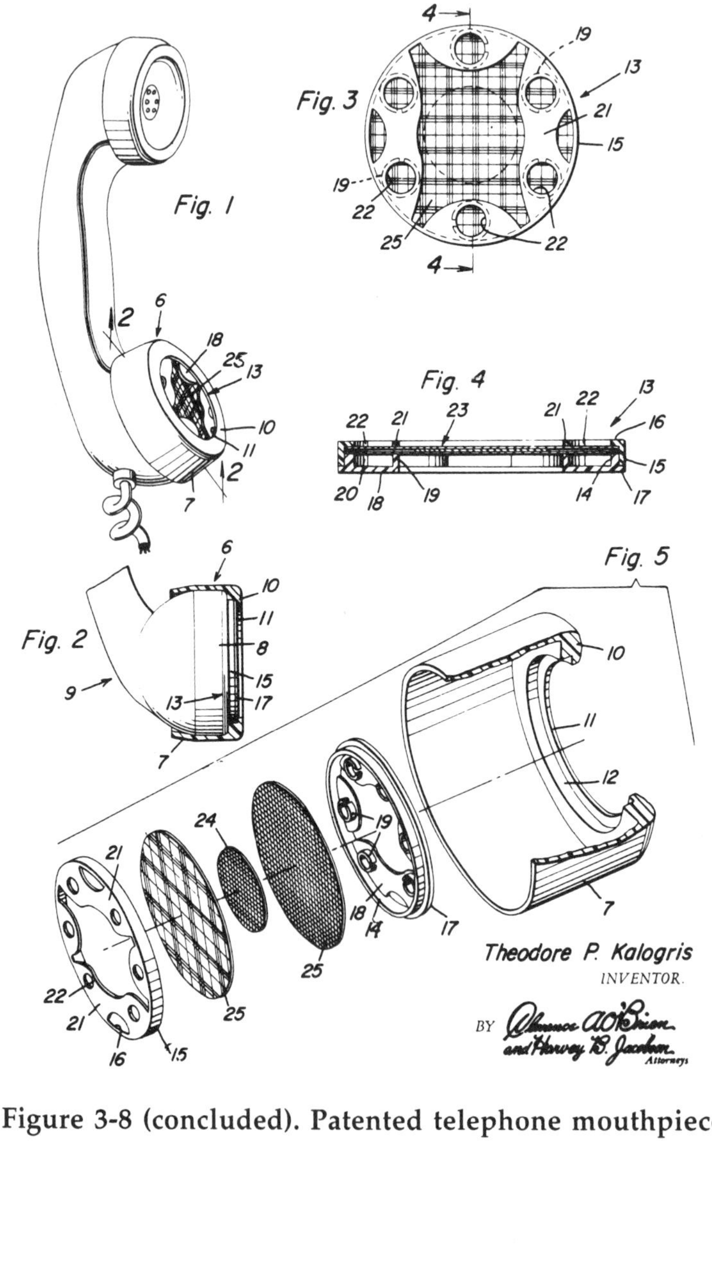

Figure 3-8 (concluded). Patented telephone mouthpiece.

Chapter 4

How to File A Patent Application

DECIDING TO FILE AN APPLICATION

Before deciding to file an application the inventor should have determined that the invention is indeed patentable. In addition, a search of previous patents should indicate that the new invention does not infringe on prior inventions. At this stage a decision should be made as to what the inventor will do with the product. If it is decided to sell or license the product, a patent is a must.

The reason why so many people hesitate in filing for a patent application is the cost for the patent attorney, which may run in excess of $5,000. In many cases, the cost of the patent attorney is considered negligible due to the fact that insuring proper protection of the invention is itself extremely valuable. A good patent attorney makes the claims of the invention broad enough to insure maximum protection while not infringing on past inventions. On the other hand, many individuals have filed their own patent applications successfully at a fraction of the cost of a patent attorney. The author does not wish to imply that everyone should file their own patent applications; rather, to emphasize that this option is available and with proper research of the invention, it is an alternative which should not be overlooked. Once a patent application is filed by either the inventor or patent attorney, the term "Patent Pending" or "Patent Applied for" can be used on the manufactured goods as a warning to competitors that an application is pending and that a patent may be issued. This warning serves several purposes, namely deterring competitors from marketing an identical product and giving a favorable impression in advertising.

THE PATENT APPLICATION

A patent application must include the following:

1) Two copies of a specification which includes both a description of the invention and the claims which are being made

2) An oath or declaration of inventorship

3) Drawing(s) of the invention, if required

4) The appropriate filing fee

5) Information Disclosure Statement (Optional)

The Specification

The specification must describe the invention fully, clearly, concisely and precisely and describe how the invention differs from pertinent prior art, including prior expired and unexpired patents. It must describe completely the process, machine, manufacture, composition of matter, or improvements invented; and the invention's principle mode of operation (if applicable). The specification should be in the following format and include the following information. All of the sections identified are illustrated in Figure 4-1.

1. *Title of the invention*

The title should be short and specific and appear as a heading on the first page of the specification. As an alternative, a more elaborate introduction can be used which states not only the title but also the name, citizenship and residence of the applicant.

2. *Abstract*

The abstract of the disclosure is a brief (four or five sentences) description of the invention which will ultimately appear on the published patent, allowing someone to get a quick and concise idea of what the invention is. The abstract is usually typed on a separate sheet and placed as the last page of the patent application.

3. *Reference to related applications*

A reference to any pending patent applications which are related to the invention which is described and claimed in the current application must be stated in this section. This is complicated because there are several types of patent applications which can be related. A patent attorney can help determine whether any related applications exist and identify what type of application it is.

4. *Background of the invention*

The background of the invention briefly states the general and specific file into which your invention falls. It also deals with prior art and the problems and deficiencies that existed. This description is important since it establishes the situation which existed in the field at the time the invention was made. There are three ways in which devices can differ, namely: structure, operation and result. A device which differs in any one of these ways is patentable. If it differs in all three ways, then it is even better.

5. *Summary of the Invention*

The summary allows those skilled in the art to determine what the invention is. It should include a description of the function of the invention and the desired results. In contrast to the abstract of the disclosure (explained above), which is limited to a few sentences, the summary can be of any length.

6. *Brief description of the drawings*

Drawings must be provided when possible to help the reader understand the invention. As a practical matter, all inventions require drawings.

7. *Detailed description of the invention*

This part of the specification must fully describe the invention, including how to make and use the invention. It must be written in clear, concise and exact terms so that a person skilled in the art, or the technology with which the invention is most nearly connected, can make and use the invention. This description must also refer to the different views and parts of the drawings if a drawing is required.

The patent specification should be sufficiently descriptive to enable someone "skilled in the relevant art" to make and use the invention. In other words, it must be written in terms sufficiently full, clear, concise and exact to enable any person skilled in the art to which the invention pertains to be able to make and use the invention. The specification must also describe the "best mode," or best method, known to the inventor for carrying out the invention.

8. *Claim or claims*

The claims section of the patent defines the metes and bounds of the patent owner's exclusive rights during the life of the patent. Every patent application must have at least one or more claims that particularly and distinctly define the subject matter that the inventor regards as his invention. A claim is easily recognized because it is the object of a single sentence that usually begins with "I claim" or "What is claimed." Claims which stand alone, i.e., have no reference to another claim, are called "independent claims" and claims which refer to only one prior claim are called "dependent claims" and are used to "narrow" the claim on some portion of the invention that was defined by the claim it refers to.

The claims must be drawn to cover that which the inventor has actually invented and nothing more. The claim is dependent upon the specification and must agree with it, and its meaning will be construed in that light. The number of claims is not limited except that each claim should preferably read upon every form of the illustrated invention. The wording and the number of claims depend upon the characteristics and the scope of the invention and the advancement of the art related to it. In some instances one short claim is adequate. It is common, though, to find patents with 20 or more claims. Usually the complexity of the invention determines the number of claims that are required. The object of the patent writer is to give the claims the greatest range of scope from broad to narrow. The broader a claim, the more valuable it usually is, but it is more likely to be invalidated by the Patent Examiner. The second line of defense is the form of a narrower claim. Assuming one or more of the broadest claims are invalidated, protection can still be afforded by the narrower claims remaining. Thus, the value of the patent may be lessened but it will still be granted.

This "narrowing" of dependent claims is best explained by referring to Patent 3,238,313 AN ANTISEPTIC TELEPHONE MOUTHPIECE GUARD, Figure 3-8 in Chapter 3. In this patent the first claim indicates what is new as:

> "An antiseptic telephone mouthpiece guard comprising a filter unit, said unit including separable telescopic inner and outer rings in an antiseptic disc, removable mounting between said rings, and means for mounting said unit on a telephone mouthpiece, said unit further including plates in one of the rings, and rings on the inner sides of said plates providing pockets for the reception of the deodorizing and disinfecting agent, said antiseptic disc forming a closure for the pockets."

When this claim is analyzed in detail, it is noticed that there are seven components which comprise the antiseptic telephone mouthpiece: inner rings, outer rings, antiseptic disc, means for mounting said unit, plates, rings and pockets. The second claim of this patent reads as follows:

> "An antiseptic telephone mouthpiece guard comprising a filter unit, said unit including separable telescopic inner and outer rings, an antiseptic disc removably mounted between said rings, and means for mounting said unit on a telephone mouthpiece, said unit further including plates in one of the rings and rings on the inner sides of said plates providing pockets for the reception of the deodorizing and disinfecting agent, said unit still further including plates on the other ring opposed to the first named plates and having openings therein aligned with the pockets, said disc being interposed between the first and second-named plates and forming a foraminous closure for the pockets."

Thus, claim 2 is narrower than claim 1, and is referred to by attorneys as "more specifically covering the invention" than claim 1, since claim 2 says that infringement would take place, not just because they are using an antiseptic telephone mouthpiece, but one that included "plates in one of the rings and rings on the inner sides of said plates providing pockets of reception of the deodorizing and disinfecting agent, said unit still further including plates on the other ring opposed to the first named plates and having openings therein aligned with the pockets, said disc being interposed between the first and second-named plates and forming a foraminous closure for the pockets."

Oath or Declaration

The oath or declaration of inventorship must state that the inventor has read and understands specification and claims; that the inventor believes himself to be the original, first and sole (or joint) inventor of the invention; and that the inventor acknowledges the duty to disclose to the Patent and Trademark Office information which is pertinent to the examination of the patent application.

Drawings

Drawings are required in most patent applications where a drawing is necessary for understanding the subject matter. The individual parts of the drawings must be referenced by letters and numerals must comply with specific requirements regarding paper size, the type of paper, margins and other details.

Filing Fee

The basic filing fee for a U.S. patent application is currently $750. For small entities (those having fewer than 500 employees), this fee is reduced to $375.

Information Disclosure Statement

There is no legal requirement for anyone to conduct a search to determine whether there is pertinent prior art, but if the inventor and the others are aware of such prior art they must disclose it to the Patent and Trademark Office. For concise reporting of such prior art, applicants are encouraged to include with the patent application a separate information disclosure statement, even though it is not required by law.

Figure 4-1. Analysis of a specification.

United States Patent [19] [11] **3,981,457**
Weist [45] **Sept. 21, 1976**

Figure 4-1 is a typical specification. Figures have been deleted for simplification. The various portions of a specification are indicated.

TITLE

[54] PLASTIC COMMINUTING METHOD AND APPARATUS
[75] Inventor: **Herman C. Weist,** Louisville, Ky.
[73] Assignee: **Weist Industries, Inc.,** Louisville, Ky.
[22] Filed: **Dec. 24, 1974**
[21] Appl. No.: **536,154**
[52] **U.S. Cl.** **241/46.17**
[51] **Int. Cl.**^{2} **B02C 23/36**
[58] **Field of Search** .... 241/16, 17, 21, 27, 241/38, 46 R, 46 B, 46.11, 46.17, 60

REFERENCES

56		References Cited	
		UNITED STATES PATENTS	
2,945,634	7/1960	Beck et al.	241/46.17 X
3,510,067	5/1970	Beck et al.	241/17
3,685,748	8/1972	Beck et al.	241/41
3,717,307	2/1973	Beck	241/46.11
3,833,178	9/1974	Beck	241/46.11

Primary Examiner—Granville Y. Custer, Jr.
Attorney, Agent, or Firm—Edward M. Steutermann

ABSTRACT

[57] **ABSTRACT**

The present invention provides a method and apparatus for comminuting selected plastic materials, where the apparatus can include a generally cylindrical receptacle means adapted to receive a selected quantity of a plastic material, to be comminuted where the receptacle has a generally vertical axis and an inlet and an outlet; cutter blade means disposed for rotation within said receptacle about an axis parallel to the vertical axis of the receptacle and in a generally horizontal plane, fixed cutter blade means carried on the sides of the receptacle means in spaced relation about the rotatable cutter blade means; fluid supply means to selectively supply a selected quantity of fluid to the receptacle; exhaust means for emission of vapor from the receptacle; lid means to selectively close a portion of the inlet to the receptacle; agitator means to selectively agitate a portion of the plastic materials received by the receptacle to urge the plastic material toward the cutter blade means; motive means for rotating the rotatable cutter blade means; power indicator means to indicate the amount of power supplied to the motive means; product exhaust means outlet means communicating with receptacle outlet for emission of comminuted plastic material from the receptacle; and fluid supply control means to actuate the fluid supply means to selectively supply fluid to the receptacle for a selected period of time; and emission control means to open the product exhaust means for a selected period of time for emission of comminuted plastic from the receptacle.

1 Claim, 3 Drawing Figures

Figure 4-1 continued.

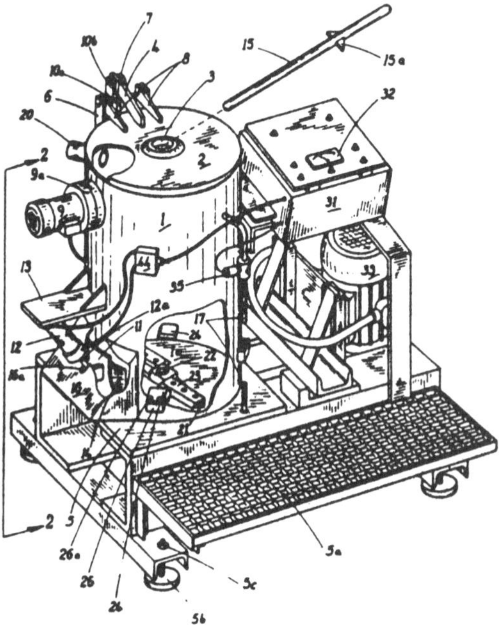

1

PLASTIC COMMINUTING METHOD AND APPARATUS

BACKGROUND OF THE INVENTION

BACK-GROUND OF THE INVEN-TION

Petroleum stock shortages and resulting increased prices enhance the attractiveness of methods and devices for reprocessing previously discarded plastic materials for use in manufacturing operations such as film waste, waste from extruding or waste and refuse from sealing and printing operations.

Various methods and apparatus as disclosed in U.S. Pat. Nos. 3,510,067 and 3,685,748 have been provided for reprocessing plastic materials, for example polyethylene, by comminution of the material and subsequent cooling of the material by fluid addition.

However, such prior apparatus does not provide means to sequentially control the steps of the processing operation to improve the efficiency of the operation by reducing processing time for multiple batches.

Figure 4-1 continued.

BACKGROUND

More specifically, such previous apparatus have not recognized certain advantages of controlling the extent of comminution of plastic material prior to addition of the fluid or the treatment of the plastic material subsequent to comminution and particularly the processes and sequences necessary after introduction of the fluid.

Likewise, such previous apparatus has not recognized the benefits of agitation of the plastic material during the comminution to maintain a constant power input to the rotating blade during the comminution cycle.

SUMMARY OF THE INVENTION

SUMMARY

The present invention provides an apparatus for comminution of selected plastic materials where it is recognized that timing and sequential operation is important to improve efficiency of operation as well as overall productivity for multiple batches and that agitation of the plastic material charged to the apparatus is important to optimize efficiency of the apparatus and method.

In accordance with another feature of the present invention it is also recognized that addition of a fluid to the comminuted plastic mass should be accomplished at a specific time during the comminution cycle characterized when the comminuted plastic particles form a toroidal shaped mass flowing around the rotating blade where the toroid of particulate material rotates around the periphery of the receptacle at a speed less than the speed of the rotating blade and can be introduced when portions of the toroidal mass begin to break away and fall into the vortex formed by the rotating blade.

Various other features of the method and apparatus of the present invention will become obvious to those skilled in the art upon reading the disclosure set forth hereinafter.

BRIEF DESCRIPTION OF THE DRAWINGS

BRIEF DESCRIPTION OF DRAWINGS

Referring now to the drawings which only show one example of apparatus in accordance with the present invention used in the description of method in accordance with the present invention.

FIG. 1 is an illustration in perspective partly in section of an apparatus in accordance with the present invention.

FIG. 2 is a view taken along line 2 - 2 of FIG. 1, with the lid open, and

Figure 4-1 continued.

2

BRIEF DESCRIPTION OF DRAWINGS (cont'd)

FIG. 3 is a schematic diagram illustrating a control circuit for use in the apparatus shown in FIG. 1 and within the scope of the present invention.

DETAILED DESCRIPTION

Referring to FIG. 1, the example of one apparatus in accordance with the present invention shown includes a generally cylindrical receptacle 1 which can be supported by a base 5, having a walkway 5*a* where receptacle 1 can be provided with a lid 2 which can be pivotably fastened to receptacle 1 by hinge means described hereinafter to be opened as shown in FIG. 2 provide access to receptacle 1.

Base 5 can be supported on shock absorbing feet 5*b* secured by cooperative bolt means 5*c* to provide means to level the apparatus and to dampen vibration of the apparatus in operation.

Receptacle 1 includes a bottom 1*a*. An observation port 3 is provided in lid 2, adapted to receive a probe 15 which can be used to agitate the plastic material charged to receptacle 1. Wings 15*a* are provided on probe 15 to engage the periphery of port 3 and are advantageously located on probe 15 to prevent the end of the probe from contacting rotating blades 23 and as described hereinafter.

A hinge arrangement, as shown, can be provided to open lid 2. In the example shown spaced parallel arms 8 are carried by lid 2 to be received between spaced generally parallel arms 6 carried by receptacle 1. A shaft 7 is provided to be received by cooperative apertures in arms 6 and 8 to provide pivot means for lid 2. A second arm 4 is carried by lid 2 to extend over the edge of lid 2 and connected to a pivotable connector 10*b* carried by the operator shaft 10*a* of an air cylinder 10 shown in FIG. 2. Operation of cylinder 10 is controlled by air supplied through solenoid air control valve 15 to cylinder 10 and, it will be understood as described hereinafter that air cylinder 10 and arm 4 open and close lid 2.

A blower 9*a*, powered by a motor 9, is provided to communicate with receptacle 1 to selectively withdraw vapor from receptacle 1. Operation of blower 9 is controlled by the control circuit as described hereinafter and located in control box 31.

A rotating arm 21 is provided in the bottom of receptacle 1 and carried for rotation in a generally horizontal plane by shaft 22 where shaft 22 is disposed generally parallel to the longitudinal axis of receptacle 1. Arm 21 is adapted to carry blades 23, which

Figure 4-1 continued.

DETAILED DESCRIPTION

can be double edged so that if one side of the blade becomes dull the blade can be reversed to provide a new edge. In the example shown arm 21 is adapted for rotation in a clockwise direction as viewed from the top of receptacle 1.

Shaft 22 is driven by an electric motor 33, for example a 3 phase 240 volt AC motor, by cooperative drive belt and pulley means as shown in FIG. 2 and further described hereinafter.

Cooperative fixed blades 24 are provided in spaced relation around the periphery of receptacle 1 and can be located a selected distance above or below the plane of rotation of blades 23. In the example shown fixed blades 24 are located a selected distance above rotating blades 23 so that plastic material charged to receptacle is comminuted between rotating blades 23 and fixed blades 24.

In the example shown blades 24 extend through cooperative openings provided in the sides of receptacle 1 and are retained by bolts 26*a* secured to lugs 26 carried on the outside of receptacle 1, as shown.

7

DETAILED DESCRIPTION

of the recognition that frequent repetition of the processing cycle permits the heat retained in the apparatus to advantageously be utilized in improving the efficiency of subsequent cycles. By the method provided by the present invention, it has been found that the bulk density of the product can be controlled and that reprocessed plastic material having a bulk density in excess of 60% of the bulk density of virgin material can be provided and that the reprocessed material provided can be used with or without blending with virgin material.

BROADENING PARAGRAPH

It will be recognized that the foregoing is but one example of an apparatus and method with the scope of the present invention and that various other modifications will occur to those skilled in the art upon reading the disclosure set forth hereinbefore:

CLAIMS

The invention claimed is:

1. An apparatus for comminuting a plastic material including: generally cylindrical receptacle means having a vertical axis, a generally horizontal bottom, an inlet opening to receive selected plastic, and outlet opening means located in the bottom portion of

Figure 4-1 continued.

8

CLAIMS

said receptacle for emission of comminuted plastic material; lid means to selectively close said inlet; outlet cover means to selectively cover said outlet means; rotatable cutter blade means adapted for rotation about an axis generally parallel to said vertical axis and in a generally horizontal plane within said receptacle and in spaced relation from said receptacle bottom; fixed cutter blade means carried on the periphery of said receptacle means in spaced relation from said rotating cutter blade means; fluid supply means selectively inject a selected fluid into said receptacle means; a generally vertical rotatable shaft means carrying said rotatable cutter blade means; bearing means located beneath and adjacent said receptacle bottom to receive said shaft means for rotation therein, and bearing cooling means to receive heat from said bearing means, where said bearing cooling means includes means to receive said fluid to be admitted to said receptacle in heat transfer relation prior to supplying said fluid to said receptacle means.

* * * * *

Figure 4-1 concluded.

Chapter 5

What Happens to the Patent Application?

Once the patent application is filed, the Commissioner initiates an examination of the application for the alleged new invention, and if on such examination it appears that the applicant is entitled to a patent under law, the Commissioner shall then issue a patent.

The examination is conducted in accordance with the rules of practice of the United States Patent Office and the Patent Office publication entitled "A Manual of Patent Examining Procedures." Once the patent is filed, the application is assigned to an examining division of the Patent Office and then to a chief or primary examiner in that division.

Applications are reviewed in sequential order based on time received. The initial check of the application ascertains whether the application is proper both in the form and content. The oath is also reviewed as to its completeness and whether it was properly executed by an official authorized to administer oaths, such as a notary public. Additional formalities are required for papers which have been executed outside the United States. In particular, the Examiner looks to see that all parts, including the summary of the invention, description of drawings, the detailed description, the mode of operation and claims, are all included in the specification. Next the Examiner makes a thorough investigation of prior patents and publications available to him. The purpose of this is to determine whether or not the invention is indeed patentable. The initial steps are done by an Assistant Examiner. The findings are reported to the Chief or to the Assistant Chief of the division and after review a letter or office action is prepared and mailed to the applicant or his attorney if there is one.

THE FLOW OF A PATENT APPLICATION

Figure 5-1 depicts the flow of a patent application from the initial filing to the end of the line. Although an inventor is permitted to act as his own attorney in patent proceedings, it is strongly recommended that an inventor have a skilled, registered patent lawyer or agent assist in the prosecution of the application. The rules for correctly prosecuting patent applications are numerous, detailed and complex.

The First Office Action

Usually within 6 to 18 months after a patent application is filed, the patent examiner will send to the applicant or the applicant's patent attorney the first written notification of his decision. This is the patent examiner's first "Official Action." If you're lucky he will write that the application is allowed, i.e., the claims are patentable. Then, upon payment of the appropriate fees, the patent is issued.

Unfortunately, more often than not, the examiner's first Official Action will say that some or all of the claims have been rejected (i.e., the claims do not meet the legal requirements for patentability). The examiner will reject all of the claims if he considers the invention to be totally unpatentable. If he considers only certain claims to be unpatentable, he will reject only those claims and indicate that the other claims are, or could be with appropriate amendment, patentable. The first Official Action will give reasons for the rejections, and will include information about prior art which may help you decide if you should continue with the prosecution of the application.

Procedures for a Patent Application

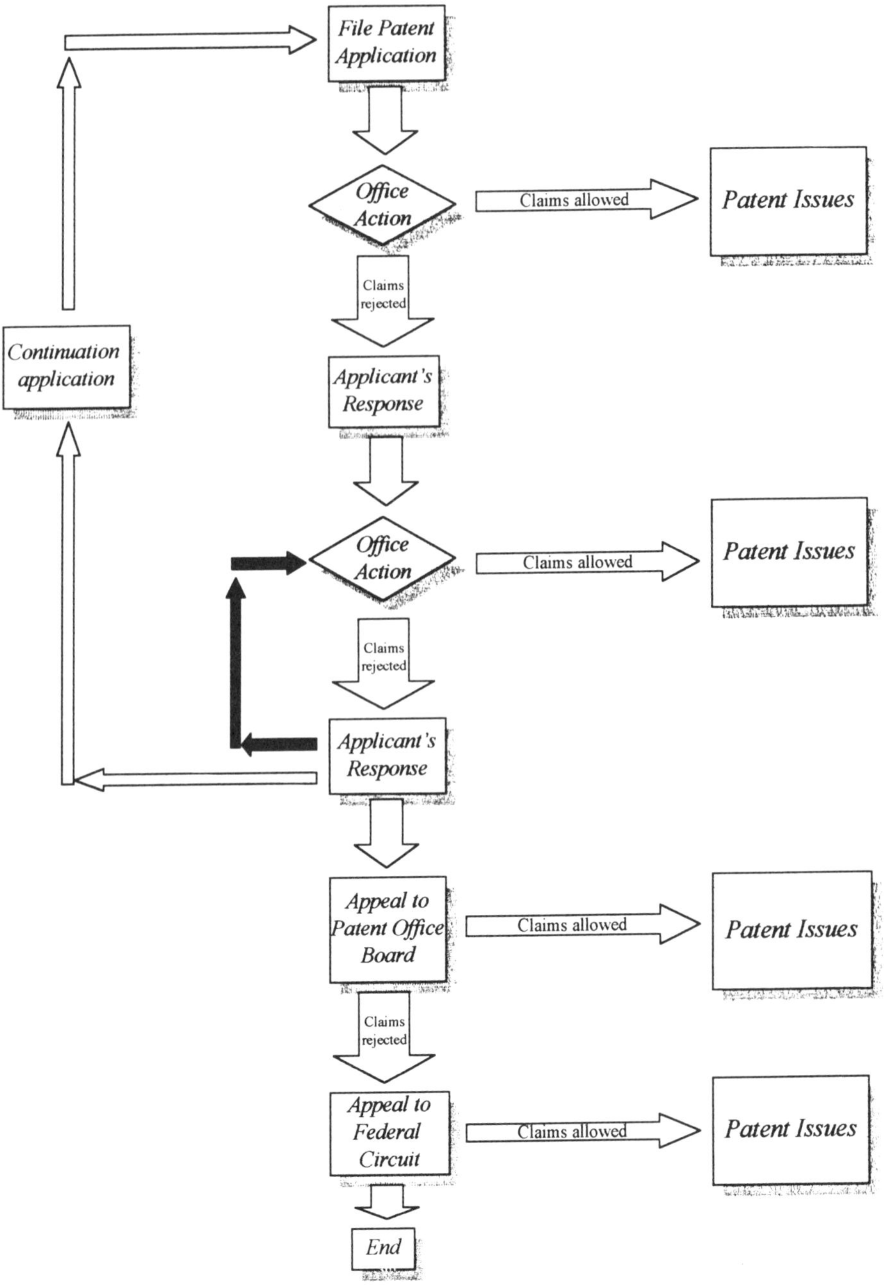

Figure 5-1. Procedures for a patent application.

Response to First Office Action

If the Examiner's first Official Action is negative in any respect, and you still want a patent, a written request must be sent to the Patent and Trademark Office asking that the application be reconsidered.

The applicant has six months (usually shortened to three months) to reply to the Patent Office actions. If the applicant decides the Examiner is correct, the applicant or his patent attorney may revise the claims or include new claims. If on the other hand the applicant believes the Examiner is wrong, a request for reconsideration can be asked for. Another method is to amend the claims and point out the supposed errors in the Examiner's actions, such as the novelty of the claims presented in view of the state of the art at the time of disclosure. The applicant must also show that the amendments avoid the objections of the Examiner. The Patent Office rules govern in preparing amendments. In preparing a reply, a discussion of the differences between the structure of the prior patents and the applicant's structure and the advantages of the differences of the applicant's structure and the shortcomings of the structure previously available can all be used to convince the Examiner that the claim is valid.

As an alternative, you can request an interview with the Examiner either by telephone or in person. If you want to interview the Examiner in person, then the interviews must be in the Examiner's room during office hours and at times convenient to the Examiner. It is not possible to have an interview before the first patent action. It is a must that the applicant or his attorney be adequately prepared with the facts and be familiar with all references cited by the Examiner. If a demonstration model exists, it is very helpful to bring this to the interview in order to show in three dimensions what was described in writing or shown in the drawings. It is possible to have an interview after final rejection as a means to obtain an allowance of claims.

It is possible that through the interview the Examiner may agree to discuss proposed claims and may also indicate that he will accept a claim or limited number of claims in the form discussed. The Examiner may also indicate his desire to search further due to new problems which have arisen in the course of the discussion. The length of the interview is usually an hour or less. The purpose of the interview is to persuade the Examiner through proper aids and understanding to appreciate the invention. At the conclusion of the interview an amendment should be made which covers the changes and revisions discussed at the interview. This practice is in conformity with the Rules of Practice which require all business with the Patent Office to be transacted in writing.

The Second Office Action

After you respond to the first Office Action, the Patent and Trademark Office will examine the application a second time and issue a second Official Action. If the claims are now found to be patentable, you only need to pay the fee and the patent will be issued. However, if the examiner rejects the claims, the examiner will probably make the second Official Action the "final rejection," which leaves you only with the option of requesting reconsideration or appealing.

Requesting Another Examination

If the claims have been rejected twice (i.e., the final rejection) you can request that the examiner review the application again, but the examiner is not legally obligated to do so; and even if he does agree to review the application, changes to the claims are not usually permitted. This means that to have any possibility of success, you must either eliminate each rejected claim or present a strong argument as to why the claim should be allowed. If at the end of your presentation the examiner feels that enough evidence exists so that the invention may be patentable, he will reexamine the application.

Appeal

You start the appeal by filing your position and paying the required fee. This gets you a review of the facts and an oral hearing by the Board of Appeals in the Patent and Trademark Office. The Board of Appeals is made up of the Commissioner of Patents and Trademarks, the Assistant Commissioners, and no more than 15 examiners-in-chief. However, each appeal is normally heard by an appeals board composed of only three individuals.

The patent will be allowable if you convince the Board of Appeals that the invention is patentable. However, if they reject your application and you do not wish to abandon your application you have the choice of appealing to the Court of Appeals for the Federal Circuit. If you decide to appeal to the Court of Appeals for the Federal Circuit, the court will review the Patent and Trademark Office records and then may uphold or reverse the action taken by the Board of Appeals. If the Court of Appeals accepts the application, you will be granted a patent. However, if they reject your application, your only alternatives are to give up or take your case to the United States Supreme Court (good luck)!

THE GRANTING OF A PATENT

If the applicant has endured the process of filing an application, survived any interferences, and succeeded in any appeals, a patent will be granted. At this point the Patent Office sends the applicant a notice of allowance. The applicant has three months to pay the final fee and correct any informalities. After the final fee is paid a patent is issued in about five to six months. On that day the patent, including the grant with its seal and ribbon, is mailed to the applicant (refer to Figure 5-2). At this time the *Official Gazette* comes out with one claim and one figure of the drawing of the patent and the Patent Office mails copies of the patent to patent subscribers, and shelves are stocked in the search room with copies of the patent. Copies are also placed on sale and mailed to libraries who subscribe to the complete files.

THE RIGHTS OF A PATENT OWNER AND PRESUMPTION OF VALIDITY

Once the patent issues, the patent protection lasts 20 years from the date of filing the application and gives the patent owner

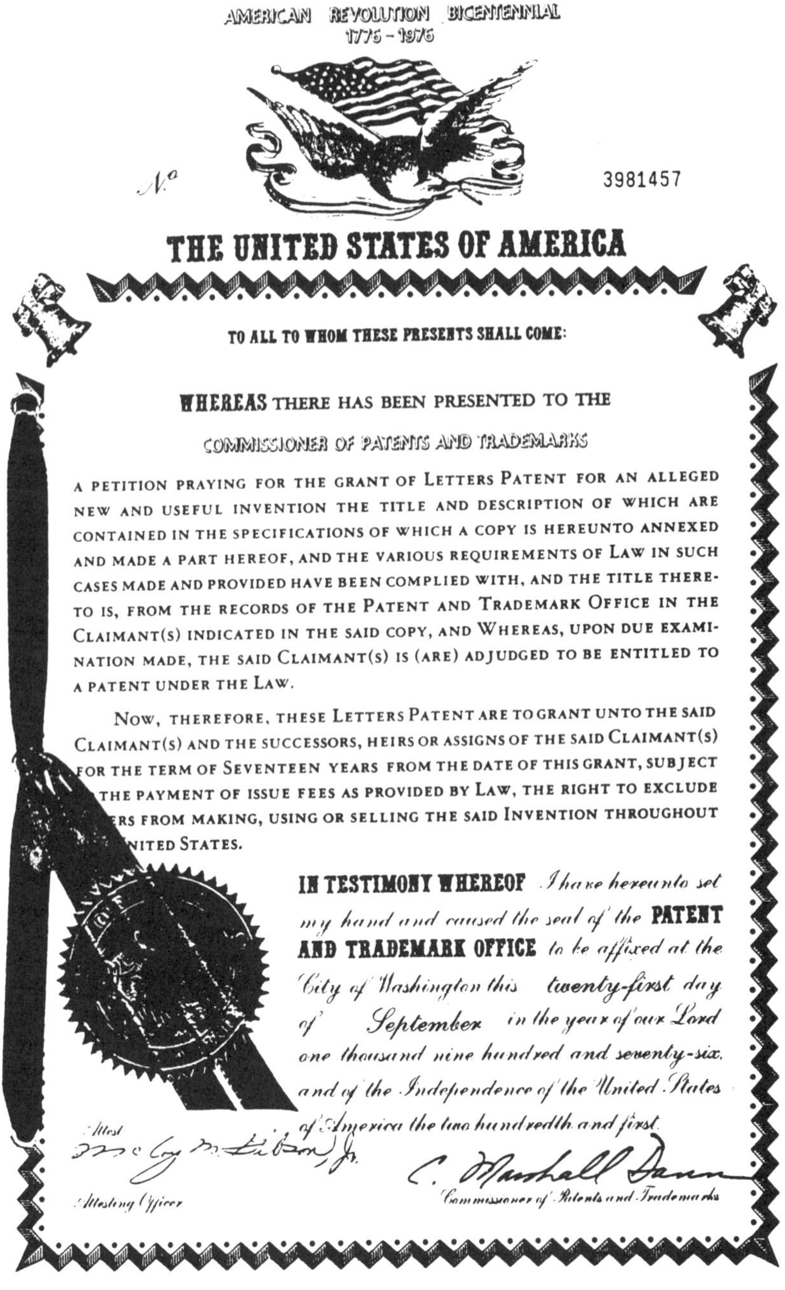

AMERICAN REVOLUTION BICENTENNIAL
1776 - 1976

No 3981457

THE UNITED STATES OF AMERICA

TO ALL TO WHOM THESE PRESENTS SHALL COME:

WHEREAS THERE HAS BEEN PRESENTED TO THE

COMMISSIONER OF PATENTS AND TRADEMARKS

A PETITION PRAYING FOR THE GRANT OF LETTERS PATENT FOR AN ALLEGED NEW AND USEFUL INVENTION THE TITLE AND DESCRIPTION OF WHICH ARE CONTAINED IN THE SPECIFICATIONS OF WHICH A COPY IS HEREUNTO ANNEXED AND MADE A PART HEREOF, AND THE VARIOUS REQUIREMENTS OF LAW IN SUCH CASES MADE AND PROVIDED HAVE BEEN COMPLIED WITH, AND THE TITLE THERETO IS, FROM THE RECORDS OF THE PATENT AND TRADEMARK OFFICE IN THE CLAIMANT(S) INDICATED IN THE SAID COPY, AND WHEREAS, UPON DUE EXAMINATION MADE, THE SAID CLAIMANT(S) IS (ARE) ADJUDGED TO BE ENTITLED TO A PATENT UNDER THE LAW.

NOW, THEREFORE, THESE LETTERS PATENT ARE TO GRANT UNTO THE SAID CLAIMANT(S) AND THE SUCCESSORS, HEIRS OR ASSIGNS OF THE SAID CLAIMANT(S) OR THE TERM OF SEVENTEEN YEARS FROM THE DATE OF THIS GRANT, SUBJECT THE PAYMENT OF ISSUE FEES AS PROVIDED BY LAW, THE RIGHT TO EXCLUDE RS FROM MAKING, USING OR SELLING THE SAID INVENTION THROUGHOUT NITED STATES.

IN TESTIMONY WHEREOF *I have hereunto set my hand and caused the seal of the* PATENT AND TRADEMARK OFFICE *to be affixed at the City of Washington this* *twenty-first* *day of* *September* *in the year of our Lord one thousand nine hundred and* *seventy-six,* *and of the Independence of the United States of America the two hundredth and first.*

Attest
McCoy M. Gibson, Jr.
Attesting Officer

C. Marshall Dann
Commissioner of Patents and Trademarks

Figure 5-2.. The patent grant.

the exclusive right to make, use or sell the invention. One may transfer patent rights to others and such transfers are recorded in the United States Patent and Trademark Office (PTO). It is important to note that a U.S. patent only protects an invention from infringement in the United States. To obtain foreign patent protection, a patent application must be filed in each country where protection is sought.

A patent owner may sue anyone he believes has been using all or part of the patent without his consent. If a patent infringement suit arises, the alleged infringer can challenge a patent's validity by demonstrating that the grant of patent was unwarranted. Once a patent issues, however, it is presumed valid, which means that the defendant must demonstrate, by clear and convincing evidence, that the Patent Office was mistaken in allowing the patent. Thus, a court which determines that an alleged infringer has not met his burden of proving the patent invalid need not affirmatively conclude that the patent is valid to uphold its validity.

CAN AN INVENTOR REVISE HIS PATENT ONCE IT IS GRANTED?

If the inventor finds an error, he may apply for a reissue of the patent. In this case the patent is surrendered and another fee must be submitted. The applicant now petitions the Commissioner to reissue the patent with the desired changes. The altered patent is referred to as the reissued patent and it expires on the date of the original patent. The original drawings may be used in the reissue if no changes were made. A new specification is required but all unchanged portions of the old specification can be reused. It is not possible to introduce new matter. It is possible to reissue the patent with broader claims, if the reissue is applied for not more than two years after the issuance of the original patent, provided that the basis of such claims existed in the specification or drawings. Claims may be narrowed at any time during the life of a patent.

Chapter 6

Guide to Patent Infringement

WHAT IS INFRINGEMENT?

Infringement is an issue in most lawsuits involving a patent. Determining patent infringement requires determining whether someone without authority makes, uses, offers, sells or imports the patented invention within the United States during the term of the patent.

HOW DOES INFRINGEMENT WORK?

For the inventor to collect damages for the infringement, the patent owner must show that fair warning was given and that the infringer continued nevertheless. Warnings can be given in two ways: either communicating directly with the infringer or by marking the patented article. Patented articles can be marked with the word "patented" followed by the patent(s) number(s). All notifications other than marking articles with the word "Patented" should be in writing. Care should be given to the wording in the notification to the potential infringer. Too harsh a statement may cause the potential infringer to file a Declaratory Judgement Suit. This suit gives the potential infringer the right to go to court to challenge the validity of the patent in question. Remember that even though a patent was granted it can be found to be invalid at a later date.

It is important to note that an infringement involves not only the person that manufactures the item but also the seller or user, who can also be liable for infringement. Therefore manufacturers of any new product are often required to guarantee to their dealers and distributors that they will assume all risks associated with damages from infringement resulting from the sale of the new product and agree to take back any such goods that may be on hand in case notices of infringement are served.

The "Patent Pending" which is marked on patents where patents have been applied for serve as a deterrent to manufacturers who want to copy an invention because the manufacturer does not know how long it will be before the patent is granted. Hence, there is a risk that the would-be infringer will be stuck with articles that can be neither used nor sold, to say nothing of the expense of the machinery used in making them. This "Patent Pending" mark gives an effective monopoly on the item even though a patent has not been granted.

A patent is infringed by a third party only if an accused device meets each and every limitation of at least one patent claim, either literally or by equivalency. A device that does not include even one limitation of a claim does not infringe that claim, and is not actionable. In any patent infringement suit, the patent is presumed, by statute, to be valid. A party found to infringe a patent is liable for damages, which may comprise lost profits or a reasonable royalty rate. The damages may be trebled in the case of willful infringement.

TWO TESTS FOR INFRINGEMENT

1. Literal Infringement

Literal infringement occurs when the exact language of a claim can be read on the conduct of a third party. To establish infringement, one must first determine the proper construction of the claim, by defining the terms used in the claim with relation to the patent specification, the prior art and the file history of the application. After the proper scope and meaning of the claims has been determined, the infringing device or method is analyzed to determine whether it exactly embodies the claimed invention.

2. Infringement under the Doctrine of Equivalents

The doctrine of equivalents allows a court to find infringement when someone steals the essence of an invention but tries to avoid liability by making a *de minimis* change. The test for equivalence is whether there are insubstantial differences between the claimed invention and the accused product or process. This test is often, but not always, satisfied by determining whether the alleged infringing product or process performs substantially the same function to achieve substantially the same result in substantially the same manner.

One major limitation on using the doctrine of equivalents which allows a patent owner a broader right to exclude than is given by the literal language of the equivalents is something called prosecution history estoppel (formerly known as file-wrapper estoppel); that is, something an applicant gave up during the patent's prosecution in order to obtain allowance of the patent cannot be recaptured by the doctrine of equivalents. For example, if the broader claims were deleted to overcome a Patent Office rejection and thereby obtain issuance of the patent, this fact can be used in an infringement case. Under these circumstances the patentee is prevented from construing that the issued claims had a broader nature based on the doctrine of equivalence. Thus, it makes no difference whether the Examiner was right or wrong with the rejection of the broader claim. By amending or substituting a narrower claim the applicant is bound by the limitation accepted. The court may look only to the claims which have been amended or substituted in determining whether the doctrine of equivalence applies and may not recognize any statements made by the applicant's council. Some courts may interpret the doctrine more liberally and then make reference to statements and arguments made by the patentee during the examination.

THE ENGINEER'S ROLE IN PATENT INFRINGEMENT

Regardless of which side of infringement situations engineers find themselves on, they play a vital and fascinating role in patent infringement cases. Often the combined talents of the patent lawyer and the engineer are required to arrive at an understanding of the scope of the claims because the law must be applied in the context of the words "to those skilled in the art," and in the context of technical literature available when the patent was applied for. The engineer-attorney team finds itself involved in the following areas:

- Determining if infringement exists;
- Assessing the validity of the patent;
- Determining the validity of the patent if licensing is considered;
- Checking for noninfringing alternatives;
- Avoiding litigation if reasonably possible;
- Evaluating arguments available to both sides.

Engineers may find themselves on the side of the company charged with infringement, or protecting a company from a would-be infringer. Engineers should make themselves generally familiar with the company's patents in the field of their work.

Since patent litigation is extremely expensive, it is avoided. There are many responses to a charge of infringement. The company which is charged can consult with its engineers to determine if indeed they are infringing upon a valid patent. If it is so determined, the company can stop infringing. A company which is charged and is convinced that their process or product does infringe upon a patent can decide to license the patent or product from the patent owner. Since infringement suits result in high legal costs and uncertainties, it is wise to consider licensing as an alternative. With the amount of new patents granted, it is quite possible that a company has infringed upon another's patents unknowingly.

Another alternative is to argue noninfringement if it is determined that there is no infringement really involved. Because of the soaring cost of litigation, many times arbitration between the patent owner and would-be infringer reaches an amicable agreement. Another alternative is to argue that the patent is not enforceable. For example, if the patent were used in a way that violates anti-trust laws, the patent—although valid—may not be enforceable. Lastly, it can be argued that the patent is invalid even though the patent was granted if it can be shown that the description of the invention does not adequately instruct those in the field how to practice that invention.

The engineer is involved in many of the areas of responses to a charge of infringement. For example, the engineer is responsible for developing information for the prior art surrounding the invention. It is possible that someone somewhere actually used the invention in public at a date earlier enough to invalidate the patent. Remember that a patent can be invalidated if the product was publicly used or was known prior to conception. The prior art search includes both foreign and domestic publications, including patents. The courts have ruled that the sale of the product, even if carried out in secret, is considered public use of the process. A patent must be applied for within one year of the first public use of a process, the sale, offer of sale, or any profit made from it. An application filed later than one year would be invalid.

If all responses to the infringement fail, the only alternative left is a trial. The patent litigation involves two principal parts: pre-trial maneuvering and the trial itself. The engineer's role is mainly confined to the pre-trial activities. After the charge of infringement the company may admit or deny infringement. If the attorney believes there is a good chance of defense, the proceedings will move to the discovery stage. The discovery stage involves interrogatories, production of documents, depositions, inspections, interparty tests and request for admissions. After the filing, both sides can begin discovery.

Let's review the pre-trial maneuverings. The plaintiff can serve a series of written questions referred to as interrogatories on the defendant. The plaintiff may require additional technical aspects of the infringer's product or process which was not available before the trial. The engineer usually advises what information can be disclosed without giving away proprietary or trade secrets. The courts will not permit the plaintiff to make a "fishing expedition" out of the discovery stage. Another maneuver is the production of documents, which involves searching company records and identifying exactly where the answers to the interrogatories can be substantiated.

The deposition stage involves taking testimony of employees involved and experts on both sides under oath in order to authenticate or challenge the documents produced. The engineer usually plays the role of a key or expert witness in these decision proceedings.

The last stage of discovery is referred to as a request for admissions. This is simply a request by either party that the other party admit certain facts from the earliest stages of discovery to be true. Many times in this stage, technical facts relating to a technical question are the issue. The interpretations placed on these facts can be supported by data from experiments. In order for the test or experiment to have an impact on the court it is suggested that representatives from both parties be present when the tests are made.

After the discovery stage, the case is ready for trial. The strategy of the trial is based on the results of the discovery, which determines which witnesses should be called and the general conduct of the case. Patent cases can be heard by the judge alone or by a jury. The trial normally consists of the witnesses' testimony related to disputed facts, introduction of documents and physical exhibits uncovered in the discovery stage.

A summary of decisions and alternatives in patent infringement situations is illustrated in Figure 6-1.

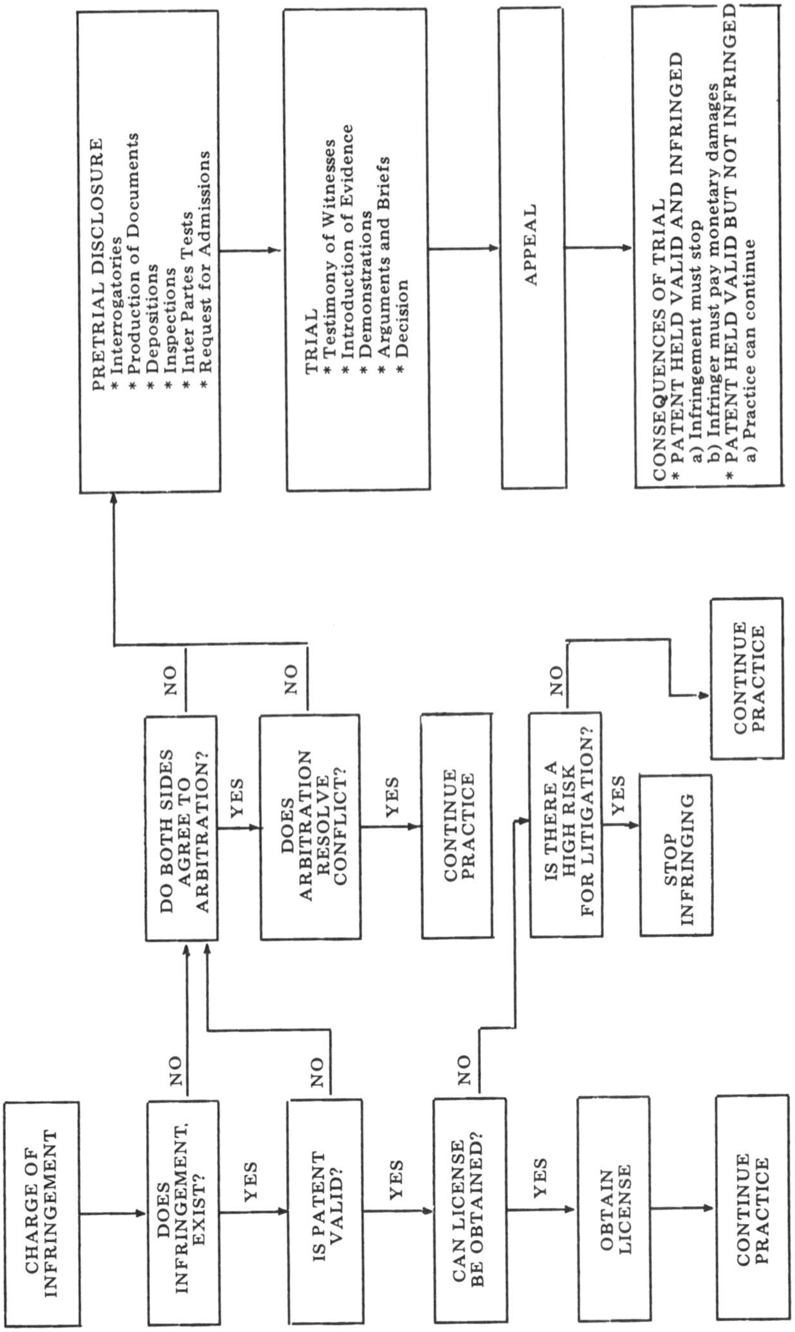

Figure 6-1. Decision matrix for patent infringement.

Chapter 7

What to Do with The Invention

Patents are personal property and therefore any legal rights, including ownership, may be transferred to others either by assignment or by license. The rights of a patent owner include the right to make the invention, the right to use the invention and the right to sell the invention. All these rights are separate legal rights and they may be assigned or licensed individually or collectively.

ASSIGNMENTS

Some or all of the ownership rights of a patent may be assigned to other individuals or companies. Once the owner of a patent assigns his rights to another party, the owner no longer owns those same rights. The most common way of assigning a patent is by selling it and there are several ways of identifying a potential buyer.

BUSINESS BROKERS

There are business associates in the field whose exclusive function is to act as brokers (not invention developers) to help the inventor sell the invention. Brokers are thought of as sales representatives who find prospects who might be interested in producing and selling the invention. They present the invention for consideration and actually handle the sales negotiations. Brokers do not get involved with patent applications or developing inventions but merely sell the invention. The United States Patent Office exercises no control over brokers and there is no professional society to establish standards. Brokers' fees vary but the commission charge is usually ten percent of the inventor's receipt. The broker acts as a screen for finding which of the more than 50,000 manufacturers may be interested in the inventor's product. Once the prospects are decided the broker submits the invention, usually through written descriptions. The documentary material presented to the prospect consists of carefully worked out descriptions of the invention and review of its possible commercial merits, a copy of the patent or the patent application, along with the drawings of the patent and the relevant material.

It is often stated that the most critical phase of selling a patent is during the final negotiations for a sale. The broker, through phone calls, letters and visits, negotiates the final sales so that the best interests of the inventor are taken into account. Once the selling process is completed, legal advisors are usually called upon to work the exact agreement of the sales. In all of these areas the inventors themselves can do the work without a broker

Approach Companies Directly

The inventor may find a willing ear from big companies if the product offered fits a need. For example, the inventors of the device that produces a stripe in toothpaste sold it to Lever Brothers. Other examples can be found with such companies as GE, IBM, DuPont, 3M, etc. The main reason these big companies bought the inventions was that the inventor provided an idea to increase their business or increase their competitive position. It is generally found that more inventions can be sold to small, specialized concerns than to the big corporations.

Other Approaches

Without the aid of a broker, the inventor usually takes a shotgun approach. Names of firms and their fields of expertise are listed in directories such as *Thomas' Register* and from mail order firms. Sending each one of the firms involved in the specialty of the invention a brochure describing the invention may arouse some interest. Another method of finding "would-be prospects" is to review trade journals for advertisements that may indicate that the product has an appeal to a company. If the inventor has a product which is similar to one that is presently manufactured, it may pay to talk to these companies first.

LICENSING

Another way patent rights may be transferred to another party is by "licensing."

A license is a contract where the patent holder, called the licensor, grants another party, called the licensee, the right to practice the invention claimed in the patent in return for considerations that the licensee gives or agrees to give the licensor. This consideration is usually in the form of payments called royalties that bear a definite relationship to the extent to which the invention is practiced. The royalty may be in the form of a single payment or series of payments. The original patent owner keeps title to the patent. Licenses are either exclusive, in which case there is only one licensee; nonexclusive, in which case the licensor may issue as many licenses as desired; or limited exclusive, in which case licenses are issued as stipulated. In nonexclusive licenses, the licensor is not bound to the number of licenses that can be issued. The modified forms of exclusive licenses are referred to as limited exclusive licenses.

DIFFERENT TYPES OF ROYALTIES

Royalties can be of three types: fixed at a certain dollar value per quantity, based on the price at which the article is sold, or based on the profit made by the manufacturer from the sales. All terms to the license are usually clearly stated in the contract. The manufacturer usually investigates the validity of the patent, potential infringement on other patents and the question of shop rights. The license agreement is usually drawn by a lawyer experienced in such matters. It is important that a lawyer review the agreement to insure that no antitrust laws are violated. The license is granted for the life of the patent. In the agreement the licensor states that he has sole ownership, that the patent is not mortgaged or otherwise encumbered, and that none of the rights granted to the inventor are assigned to anyone else by license grant or shop rights except as stated. All licenses as well as assignments should be recorded with the Patent Office. If the license issued is to be nonassignable or nontransferable to another party, it should be so stated in the license agreement. A problem which arises is that the licensee who is making and selling the article usually improves upon the patentable item during the course of its life. Unless the provisions are spelled out in the license agreement misunderstandings may arise later on. A statement such as the following is sometimes included in the license agreement:

> "The licensor states that there are no patent, or patent pending applications for any invention in the same art as the patents that are subject of the license and that patent owners agree to give the licensee the benefit of any improvements that are made"

In some situations the licensee agrees to sign all improvements made to the licensor, reserving the right to use an improvement just as if it had been included in the original contract.

License agreements should have clauses providing for automatic termination without notice in case of bankruptcy, and a clause for terminating the license in case of insolvency or default by the licensee of any of the terms of the license agreement. In order for the license agreement to be binding there should be earnest consideration that the parties are going to be paid if they accomplish what they undertake. Figure 7-1 illustrates a typical royalty agreement.

INTRODUCTION

This agreement, entered into this _ _ _ _ _ day of _ _ _ _ _ by and between _ _"X"_ _ of _ _ _ _ _, hereinafter called _ _"X"_ _ which expression shall include his executors, administrators, and assigns, where the text so requires or admits, and the _ _"Y"_ _ Company, of _ _ _ _ _ a corporation organized under the laws of the state of _ _ _ _ _, hereinafter called _ _"Y"_ _ which expression shall include its successors and assigns, where the text so requires or admits, witnesses that-

FACTS OF OWNERSHP

In consideration of their mutual grants and undertakings, the parties hereto agree as follows:

_ _"X"_ _ is the sole owner of the following patents, which are hereinafter referred to as the _ _"X_ _ patents":

U.S. Patent No. _ _ _ _ _, dated _ _ _ _ _ for a _"Z"_

U.S. Patent No. _ _ _ _ _, dated _ _ _ _ _ for an Improvement in _ _"Z"_ _,

STATEMENT OF OWNERSHP

_ _"X"_ _ guarantees that patents Nos. _ _ _ _ _ are the only United States patents relating to _ _"Z"_ _ owned by him, and that patents Nos. _ _ _ _ _ are expressly excluded from this license; and that he has no application pending for United States or foreign patents pertaining to _ _"Z"_ _, and agrees that any other patents, which he may now own or which he may acquire, relating to _ _"Z"_ _, or based upon any invention now owned or controlled by, or whose acquisition is now contemplated by _ _"X"_ _ shall be considered as included in the preceding paragraph.

_ _"X"_ _ warrants that he is the sole owner or all rights, title, and interests in and to the _ _"Z"_ _ patents, that he has the unrestricted right to make this agreement pertaining to them, and that there are no assignments, grants, licenses, mortgages, or other encumbrances, agreements, or shop rights, either written, oral, or implied, pertaining to any of them.

(*Continued*)

Figure 7-1. The structure of a royalty license agreement.

DEFINITIONS

The word "patent" as used herein shall include a reissue, divisional issue, division, continuation, continuation in part, or extension of such patent.

The word "application" as used herein shall include any division, revival, or renewal, reissue, continuation, or continuation in part thereof, any new application filed in place thereof, and, wherever the text will admit, any patent that may issue on any of the foregoing.

The word "affiliate" (or "affiliates") as used herein shall be construed to mean any company (or companies), firm, or person in which _ _"Y"_ _ has capital stock ownership or with which it has management relations sufficient to enable it to influence the business policy and activities of such company, firm, or person.

Any notice required by the terms of this agreement to be given by either party to the other, shall, unless otherwise herein specified, be given by sending it by registered mail, properly addressed and prepaid, in the case of _ _"X"_ _ to his last-known [address]...

THE GRANT

Subject to the terms and conditions hereinafter set forth, _ _"X"_ _ hereby grants to _ _"Y"_ _ an exclusive license throughout the United States and the Territories thereof, but not elsewhere, to make, have made, use, and vend the inventions claimed by the said _ _"X"_ _ patents, for the term ending with the expiration of that one of the said _ _"X"_ _ patents having the longest time to run, whether it be one of the _ _"X"_ _ patents, or whether it be a United States or foreign patent hereafter to be issued and to be included in this license as hereinafter provided.

EXCEPTION TO "EXCLUSIVE" RIGHTS

_ _"X"_ _ reserves to himself the right, and _ _"Y"_ _ concedes such right, to make, have made, use, and vend _ _"Z"_ _ devices claimed in the _ _"X"_ _ patents or any claim in any one of them.

/more/

Figure 7-1 (continued).

CONSIDERATIONS

_ _"Y"_ _ admits, and agrees never to contest, the validity of the _ _"X"_ _ patents listed herein.

_ _"Y"_ _ agrees not to assign this license or any part thereof, and not grant any sublicenses hereunder without first obtaining _ _"X"_ _ written consent; except that, if _ _"Y"_ _ business be sold or transferred as a whole, _ _"Y"_ _ shall have the right to assign this license to its successor without the consent of _ _"X"_ _.

_ _"X"_ _ agrees to give his written consent to the issuance of sublicenses by _ _"Y"_ _ to any or all of its affiliates, provided, however, that _ _"Y"_ _ shall request such consent in writing, stating the relations of _ _"Y"_ _ to the company for which a sublicense is requested, and, further, provided that the rate of royalty and of the terms and conditions relating thereto shall be the same in the sublicenses as herein set forth, and that all royalties accruing under any and all such sublicenses shall be payable by the sublicensees directly to _ _"X"_ _.

_ _"Y"_ _ agrees not to sell any article made under this license to any of its affiliates to which sublicenses shall not have been issued, at a lower price than such article of the same size and type shall be currently quoted by _ _"Y"_ _ and its affiliates, the express intent of this and the preceding paragraph being to prevent _ _"Y"_ _ from selling to its affiliates at prices that would reduce _ _"X"_ _ royalties.

_ _"Y"_ _ agrees to pay to _ _"X"_ _ a royalty of _ _ _ percent (_ _ _%) of the net selling price of all articles sold hereunder. The net selling price shall be construed to mean the price at which the articles shall be sold by _ _"Y"_ _ or its affiliates after all regular trade discounts have been deducted, but no allowances for prompt payment of account or any other allowances or rebates shall be deducted in computing the net selling price.

_ _"Y"_ _ shall at all times keep an accurate record of all articles sold hereunder and shall render a full and complete

/more/

Figure 7-1 (continued).

CONSIDERATIONS

statement of the same, in writing, on or before the tenth (10th) day of each calendar month, for all articles sold during the preceding calendar month and said report shall show the quantities, classes, and net selling prices of the articles sold during the term reported.

_ _"Y"_ _ agrees to pay the said royalties on or before the tenth (10th) day of each calendar month, on all articles sold hereunder during the preceding calendar month.

If _ _"Y"_ _ fails to pay the royalties as above provided, and if _ _"X"_ _ thereafter notifies _ _"Y"_ _ in writing of his intention to terminate this license thirty (30) days from the date of the said notice, _ _"X"_ _ shall have the right at the end of that time to terminate this license, provided that any royalties remain unpaid at the expiration of the thirty-day period. The said termination shall not prejudice any right of action or remedy of _ _"X"_ _ for the recovery of any money then due hereunder, nor shall it prejudice any other right of _ _"X"_ _ hereunder.

The minimum royalties to be paid by _ _"Y"_ _ to _ _"X"_ _ hereunder shall be _ _ _ _ _ hundred dollars ($_ _ _ _ _) for the first year, _ _ _ _ _ hundred dollars ($_ _ _ _ _) for the second year, and _ _ _ _ _ hundred dollars ($ _ _ _ _ _) for each year thereafter during the term of this license.

If _ _"Y"_ _ fails in any year to pay the specified minimum royalties, it shall have the right to pay _ _"X"_ _ within thirty (30) days after the expiration of the twelve-month period, the difference between the amount of royalties it actually paid and the minimum specified for that year. Amounts paid in excess of the minimum requirements for one year shall not be credited on the minimum for any subsequent year. Upon failure by _ _"Y"_ _ to pay the minimum amount within thirty (30) days after the expiration of the twelve-month period, _ _"X"_ _ shall have the right, at his option, to terminate the license upon giving thirty (30) days' prior written notice to _ _"Y"_ _. If

/more/

Figure 7-1 (continued).

CONSIDERATIONS

_ _"Y"_ _ pays the amount due within the thirty-day period, _ _"X"_ _'s option to terminate shall automatically cease until the next twelve-month period.

If _ _"X"_ _ and _ _"Y"_ _ disagree upon the interpretation of any clauses herein, or upon anything pertaining to the subject of this license not specifically covered herein, the matter of dispute shall, upon request of one party in writing given to the other, be referred to an arbitration committee composed of one member appointed by _ _"X"_ _, and one member appointed by_ _"Y"_ _, and those two shall promptly agree upon and appoint a disinterested third member. The findings of a majority of the committee reported in writing to _ _"X"_ _ and _ _"Y"_ _ and signed by at least two members shall be binding upon both parties hereto.

CONCLUDING PHRASES

In witness whereof, _ _"X"_ _ has hereunto set his hand and seal, and _ _"Y"_ _ has caused its name to be subscribed and its corporate seal affixed by its duly authorized officers at _ _ _ _ _ this _ _ _ _ _ day of _ _ _ _ _ .

_ _"X"_ _ (seal)

_ _"Y"_ _

by _ _ _ _ _

its President

Attest: _ _ _ _ _

its Secretary

(corporate seal)

County
_ _ _ _ _ _ _ _ _ _ } SS

_ _"X"_ _, and _ _"Y"_ _ to me personally known, this day appeared before me and acknowledged the execution of the foregoing instrument to be his free act and deed.

Notary Public

(Notary's Seal)

(Here appears the Notary's statutory statement about his commission.)

Figure 7-1 concluded.

STARTING YOUR OWN BUSINESS

Of the remaining methods of exploiting inventions the one for greatest profit potential is for the inventors to form their own company. Before starting a business the inventor should consider the following:

- How can capital be obtained for the business?
- Can production problems be solved to economically produce the product?
- How will the product be sold?

If an inventor has decided to create his or her own company to produce the product it may be decided that the best thing is not to patent the product and to keep the invention secret. It may be far better protection, for instance, not to publish a chemical formula but rather to produce it and protect it as a trade secret.

The biggest problem inventors face is to accurately estimate the cost of developing and marketing their new product. It is often found that an inventor seems to be on the low side of costs by a factor of three or more. There have been many success stories of inventors who started their own business to manufacture and sell a product. Few individuals become rich by royalty arrangements on their inventions. On the other hand, there have been many success stories of individuals who have produced their product from start to finish with their own company. When starting your own business don't overlook the invaluable aid given to small businesses by the Small Business Administration. Not only can inventors discuss their potential business with retired businessmen but they can also learn how to secure a loan.

Chapter 8

How to Obtain An International Patent

U.S. patent protection is enforceable only within the United States. Patents are enforceable only in the country that granted them. Foreign patent rights are typically granted on a country-by-country basis, and each country has its own rules for determining patentability, which often differ significantly from those of the U.S. Patent Office. Some of the most significant differences are enumerated below.

DIFFERENCES BETWEEN MOST FOREIGN COUNTRIES AND THE U.S.

1. "First to file" vs. "First to invent"

Most foreign countries (including Europe and Japan) award patents to the first-to-apply rather than the U.S. system of first-to-invent. Thus, in Europe and Japan, patenting is essentially a race to the patent office. A party that is even one day later than another party may lose all rights. Indeed, the application of the first party will be used as prior art to the second for all that it discloses. Thus, the second application will be patentable only if it describes some aspect of the invention that is not disclosed in the first application.

2. No Grace Period

In most foreign countries (including Europe and Japan) systems, you lose the right to patent protection unless you file a patent application before the first public disclosure for sale of the invention. In contrast, the U.S. allows a one-year grace period in which to file your U.S. application after first public disclosure.

3. Automatic Publication

In most foreign countries (including Europe and Japan) applications are published before they are issued as bona fide patents whereas in the United States patent applications are maintained in secrecy until they issue.

PARIS CONVENTION

Most of the world's industrialized countries are parties to an international treaty known as the "Paris Convention." The convention is a multinational treaty that defines a right of priority. Virtually all developed countries are signatories to the Convention (Taiwan being a notable exception). Prior to the execution of the Paris Convention, parties who wished to secure patent protection in several countries needed to file their applications in each country on the same date. The Convention establishes a priority date for each invention, and provides that any foreign filing that is completed within 12 months of the priority date will be considered as though filed on the priority date for purposes of determining conception. The priority date is the filing date of the first application by that applicant which describes the invention.

For example, if a U.S. patent application is filed on January 1, 1996, the Paris Convention will permit corresponding applications, each having an effective date of January 1, 1996, to be filed in other member countries until January 1, 1997. Assuming that you filed your U.S. application before any publication or commercialization of the invention, the Paris Convention generally gives you one year from your U.S. filing date to file your foreign patent applications. However, to obtain patent protection in countries that are not members of the Paris Convention, like Taiwan and India, you must file your patent applications in those countries prior to the first public disclosure or sale of the invention.

PATENT COOPERATION TREATY

The Patent Cooperation Treaty or "PCT" provides a two-stage examination process for applications, first at an international level and then in the individual countries from which patents are sought. The signatories include Australia, France, New Zealand, Austria, Gabon, Niger, Barbados, Germany, Norway, Belarus, Greece, Poland, Belgium, Guinea, Portugal, Benin, Hungary, Rep. of Korea, Brazil, Ireland, Romania, Bulgaria, Italy, Russian Fed., Burkina Faso, Japan, Senegal, Cameroon, Kazakhstan, Slovakia, Canada, Latvia, Spain, Central African Rep., Liechtenstein, Sri Lanka, Chad, Luxembourg, Sudan, China, Madagascar, Sweden, Congo, Malawi, Switzerland, Côte D'Ivoire, Mali, Togo, Czech Rep., Mauritania, Ukraine, Dem. People's Rep. of Korea, Monaco, United Kingdom, Denmark, Mongolia, Uzbekistan, Finland, Netherlands and Vietnam.

The basic way the PCT works is the following: an inventor files a single application in any one of the PCT countries which participates fully in the system, and at the same time specifies additional participating PCT countries in which he would like patent protection. One of the PCT patent examining offices around the world examines the application for patentability in all of the countries designated by the inventor and who take part in the system.

Once the request has been filed, the applicant has 20 months (from the priority date) to translate the application into the foreign language of designated countries and to start the actual filing of the application in such countries. The 20-month deadline can be extended to 30 months for some jurisdictions. Thus, the PCT system permits an applicant to secure the right to foreign file while deferring the cost of national filings for a considerable period. This allows someone to determine whether the invention is important enough to warrant the expense of seeking foreign patent rights.

SUMMARY OF PCT vs. PARIS CONVENTION

In short, the PCT has a potentially significant advantage over the Paris Convention: it gives you up to 18 more months to decide on what foreign patent protection is needed without having to plunge into applying for many foreign patents just to be on the safe side. The disadvantage is that it costs somewhat more to file under the PCT. While these additional costs are relatively small in each country, they can add up if you file patent applications in several countries. You and your patent lawyer need to make a conscious decision for each patent application as to whether the PCT's potentially extra 18 more months to make foreign patent filing decisions is worth the extra cost.

Chapter 9

A Compendium of Handy Working Aids

GENERAL INFORMATION AND CORRESPONDENCE

All business with the Patent Office should be transacted in writing and all letters addressed to the Commissioner of Patents, Washington, D.C., 20231. The personal attendance of applicants at the Patent Office is unnecessary.

Applicants and attorneys will be required to conduct their business with the Office with decorum and courtesy. Papers presented in violation of this requirement will be returned.

A separate letter (but not necessarily a separate envelope) should be written in relation to each distinct subject of inquiry such as assignments for recording, payment of issue fees, orders for printed copies of patents, orders for photographic copies of records and requests for other services. None of these should be included with letters responding to Office action in applications.

When a letter concerns a patent, it should state the name of the patentee, the invention, and the patent number and date.

In making inquiry concerning the status of an application, the inventor should be sure to give its serial number and filing date.

U.S. PATENT AND TRADEMARK OFFICE
Effective October 1, 1995

The U.S. Patent and Trademark Office has amended its rules of practice in patent and trademark cases, Parts 1 and 2 of Title 37, Code of Federal Regulations, to adjust certain patent and trademark fee amounts to reflect fluctuations in the Consumer Price Index (CPI).

Any fee payment due and paid on or after October 1, 1995, must be paid in the revised amount. The date of mailing indicated on a proper Certificate of Mailing or Transmission under 37 CFR 1.8 will be considered to be the date of receipt and payment in the Office.

As this fee sheet is a summary and the content of rules also may be changing, you should refer to the notice published in the *Federal Register* on August 11, 1995 in Volume 60, Number 155, pages 41018 through 41027. See also the *Official Gazette of the United States Patent and Trademark Office* of August 29, 1995. The fees, which are subject to reduction for small entities who have established status (37 CFR 1.27), are shown in a separate column. For additional information, please contact the PTO Public Service Center at (703) 308-HELP.

Description	Fee	Small Entity Fee if applicable
Patent Filing Fees		
Basic filing fee - utility	750.00	375.00
Independent claims in excess of three	78.00	39.00
Claims in excess of twenty	22.00	11.00
Multiple dependent claim	250.00	125.00
Surcharge - Late filing fee or oath or declaration	130.00	65.00
Design filing fee	310.00	155.00
Plant filing fee	510.00	255.00
Reissue filing fee	750.00	375.00
Reissue independent claims over original patent	78.00	39.00
Reissue claims in excess of 20 and over original patent	22.00	11.00
Provisional application filing fee	150.00	75.00
Surcharge - Late provisional filing fee or cover sheet	50.00	25.00
Non-English specification	130.00	

(Continued)

Figure 9-1. Schedule of patent fees.

Description	Fee	Small Entity Fee if applicable
Patent Issue Fees		
Utility issue fee	1,250.00	625.00
Design issue fee	430.00	215.00
Plant issue fee	630.00	315.00
Patent Maintenance Fees		
Applications Filed on or after December 12, 1980		
Due at 3.5 years	990.00	495.00
Due at 7.5 years	1,990.00	995.00
Due at 11.5 years	2,990.00	1,495.00
Surcharge - Late payment within 6 months	130.00	65.00
Surcharge after expiration - Late payment is unavoidable	660.00	
Surcharge after expiration - Late payment is unintentional	1,550.00	
Miscellaneous Patent Fees		
Extension of term of patent	1,060.00	
Requesting publication of SIR - Prior to examiner's action	870.00*	
Requesting publication of SIR - After examiner's action	1,740.00*	
For filing a submission after final rejection (see 37 CFR 1.129(a))	750.00	375.00
For each additional invention to be examined (see 37 CFR 1.129(b))	750.00	375.00
Certificate of correction	100.00	
For filing a request for reexamination	2,390.00	
Statutory Disclaimer	110.00	55.00

*Reduced by Basic Filing Fee Paid

REMITTANCES FROM FOREIGN COUNTRIES MUST BE PAYABLE AND IMMEDIATELY NEGOTIABLE IN THE UNITED STATES FOR THE FULL AMOUNT OF THE FEE REQUIRED.

(Continued)

Figure 9-1. Schedule of patent fees.

Description	Fee	Small Entity Fee if applicable
<u>Patent Extension Fees</u>		
Extension for response within first month	110.00	55.00
Extension for response within second month	380.00	190.00
Extension for response within third month	900.00	450.00
Extension for response within fourth month	1,400.00	700.00
<u>Patent Appeals/Interference Fees</u>		
Notice of appeal	290.00	145.00
Filing a brief in support of an appeal	290.00	145.00
Request for oral hearing	250.00	125.00
<u>Patent Petition Fees</u>		
Petitions to the Commissioner, unless otherwise specified	130.00	
Submission of an Information Disclosure Statement (§1.97(c))	220.00	
Petition to institute a public use proceeding	1,430.00	
Petition to revive unavoidably abandoned application	110.00	55.00
Petition to revive unintentionally abandoned application	1,250.00	625.00
Petitions related to provisional applications	50.00	
<u>PCT Fees - National Stage</u>		
Oath or declaration after twenty or thirty months from priority date	130.00	65.00
English translation after twenty or thirty months from priority date	130.00	
IPEA - U.S	680.00	340.00
ISA - U.S	750.00	375.00
PTO not ISA or IPEA	1,010.00	505.00
Claims meet PCT Article 33(1)-(4) - IPEA - U.S.	94.00	47.00
Claims - extra independent (over three)	78.00	39.00

(Continued)

Figure 9-1. Schedule of patent fees.

Description	Fee	Small Entity Fee if applicable
Claims - extra total (over twenty)	22.00	11.00
Claims - multiple dependent	250.00	125.00
For filing with EPO or JPO search report	880.00	440.00
<u>PCT Fees - International Stage</u>		
Transmittal fee	220.00	
PCT search fee - no U.S. application	660.00	
Supplemental search per additional invention	190.00	
PCT search- prior U.S. application	430.00	
Preliminary examination fee - ISA was the U.S.	470.00	
Preliminary examination fee - ISA not the U.S.	710.00	
Additional invention - ISA was the U.S.	140.00	
Additional invention - ISA not the U.S.	250.00	
<u>PCT Fees to WIPO</u>		
Basic fee (first thirty pages)	604.00*	
Basic supplemental fee (for each page over thirty)	12.00*	
Handling fee	185.00*	
Designation fee per country	147.00*	
<u>PCT Fee to EPO</u>		
International search	1,722.00*	

Description	Fee
<u>Patent Service Fees</u>	
Printed copy of patent w/o color, regular service	3.00
Printed copy of patent w/o color, overnight delivery to PTO box or overnight fax	6.00
Printed copy of patent w/o color, ordered via expedited mail or fax, exp. service	25.00
Printed copy of plant patent, in color	12.00

*WIPO and EPO fees subject to periodic change due to fluctuations in exchange rate. Refer to Patent Official Gazette for current amounts.

(*Continued*)

Figure 9-1. Schedule of patent fees.

Description	Fee
Copy of utility patent or SIR, with color drawings	24.00
Certified or uncertified copy of patent application as filed, regular service	15.00
Certified or uncertified copy of patent application, expedited local service	30.00
Certified or uncertified copy of patent-related file wrapper and contents	150.00
Certified or uncertified copy of document, unless otherwise provided	25.00
For assignment records, abstract of title and certification, per patent	25.00
Library Service	50.00
List of U.S. patents and SIRs in subclass	3.00
Uncertified statement re status of maintenance fee payments	10.00
Copy of non-U.S. document	25.00
Comparing and Certifying Copies, Per Document, Per Copy	25.00
Additional filing receipt, duplicate or corrected due to applicant error	25.00
Filing a Disclosure Document	10.00
Local delivery box rental, per annum	50.00
International type search report	40 00
Self-service copy charge, per page	0.25
Recording each patent assignment, agreement or other paper, per property	40.00
Publication in Official Gazette	25.00
Labor charges for services, per hour or fraction thereof	30.00
Unspecified other services	AT COST
Retaining abandoned application	130.00
Handling fee for incomplete or improper application	130.00
APS-Text terminal session time, per hour	40.00
APS-CSIR terminal session time, per hour	50.00
Patent coupons	3.00
Handling fee for withdrawal of SIR	130.00
<u>Patent Enrollment Fees</u>	
Admission to examination	310.00

(Continued)

Figure 9-1. Schedule of patent fees.

Description	Fee
Registration to practice	100.00
Reinstatement to practice	15.00
Copy of certificate of good standing	10.00
Certificate of good standing - suitable for framing	20.00
Review of decision of Director, Office of Enrollment and Discipline	130.00
Regrading of Examination	130.00

GENERAL FEES

Finance Service Fees	
Establish deposit account	10.00
Service charge for below minimum balance	25.00
Service charge for below minimum balance restricted subscription deposit account	25.00
Processing returned checks	50.00
Computer Service Fees	
Computer records	AT COST

Figure 9-2. Schedule of trademark fees.

Trademark Processing Fees	
Application for registration, per class	245.00
Filing an Amendment to Allege Use under § 1(c), per class	100.00
Filing a Statement of Use under § 1(d)(1), per class	100.00
Filing a Request for a Six-month Extension of Time for Filing a Statement of Use under § 1(d)(1), per class	100.00
Application for renewal, per class	300.00
Additional fee for late renewal, per class	100.00
Publication of mark under § 12(c), per class	100.00
Issuing new certificate of registration	100.00
Certificate of Correction, registrant's error	100.00
Filing disclaimer to registration	100.00
Filing amendment to registration	100.00

(Continued)

Figure 9-2. Schedule of trademark fees (*Continued*).

<u>Trademark Processing Fees</u>

Filing § 8 affidavit, per class	100.00
Filing § 15 affidavit, per class	100.00
Filing combined §§ 8 & 15 affidavit, per class	200.00
Petition to the Commissioner	100.00
Petition for cancellation, per class	200.00
Notice of opposition, per class	200.00
Ex parte appeal, per class	100.00
Dividing an application, per new application (file wrapper) created	100.00

<u>Trademark Service Fees</u>

Printed copy of each registered mark, regular service	3.00
Printed copy of each registered mark, overnight delivery to PTO box or overnight fax	6.00
Printed copy of each registered mark ordered via expedited mail or fax, exp. service	25.00
Certified copy of registered mark, with title and/or status, regular service	10.00
Certified copy of registered mark, with title and/or status, expedited local service	20.00
Certified or uncertified copy of trademark application as filed, regular service	15.00
Certified or uncertified copy of trademark application as filed, expedited local service	30.00
Certified or uncertified copy of trademark-related file wrapper and contents	50.00
Certified or uncertified copy of trademark document, unless otherwise provided	25.00
For assignment records, abstracts of title and certification per registration	25.00
Comparing and certifying copies, per document, per copy	25.00
Self-service copy charge, per page	0.25
Recording trademark assignment, agreement or other paper, first mark per document	40.00

(Continued)

Figure 9-2. Schedule of trademark fees (*Concluded*).

Trademark Service Fees

For second and subsequent marks in the same document	25.00
Labor charges for services, per hour or fraction thereof	30.00
Unspecified other services	AT COST
Each hour of X-SEARCH terminal session time	40.00
Trademark coupons	3.00

SYMBOLS FOR DRAFTSMEN

Graphical symbols for conventional elements may be used on patent drawings when appropriate, subject to approval by the Patent Office. The symbols and other conventional devices which follow (Figure 9-3) have been and are approved for such use.

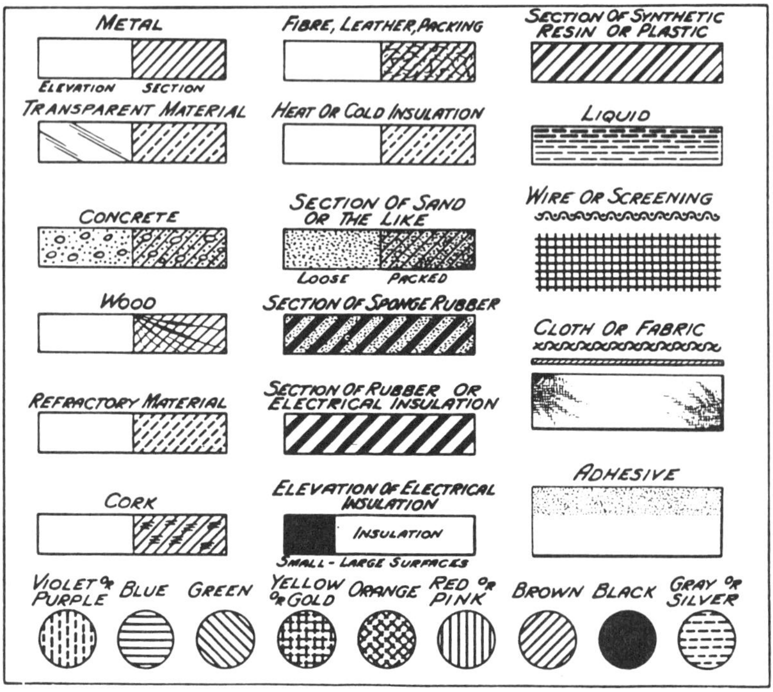

Figure 9-3. Approved graphical symbols for patent drawings.

Electrical Symbols

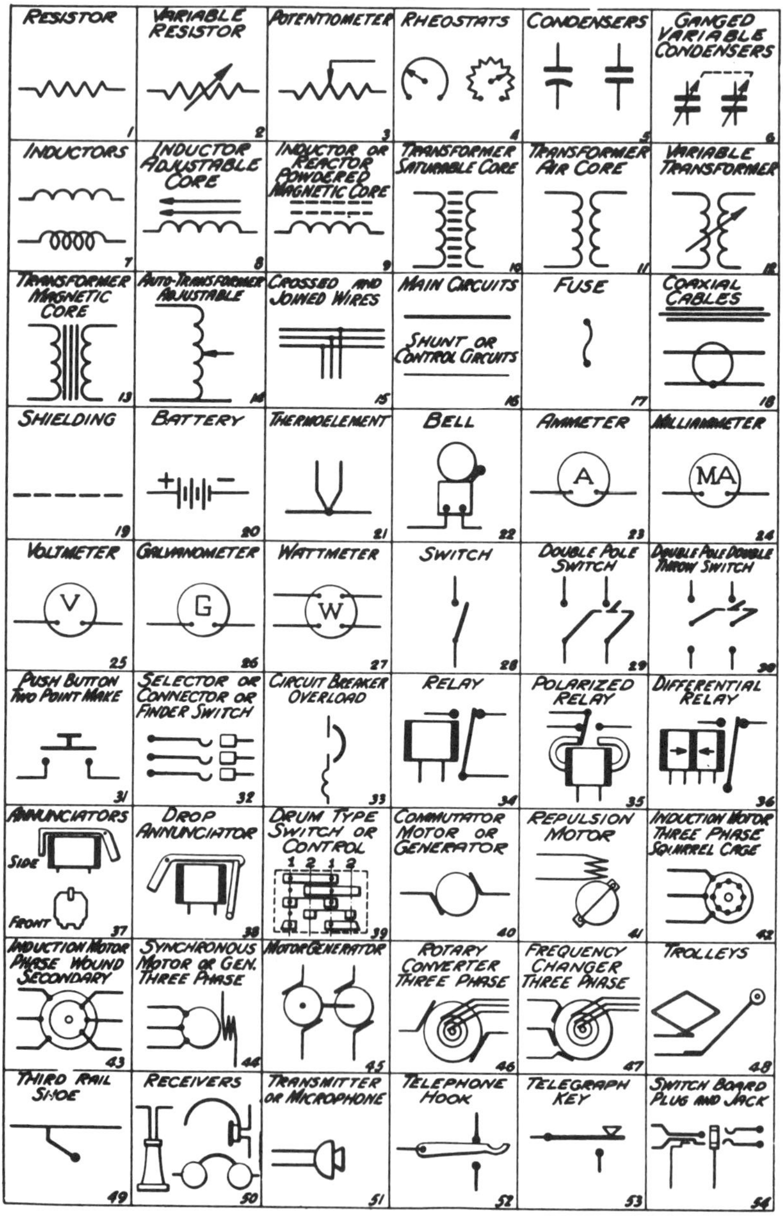

Figure 9-3 (*Continued*).
Approved graphical symbols for patent drawings.

Electrical Symbols – continued

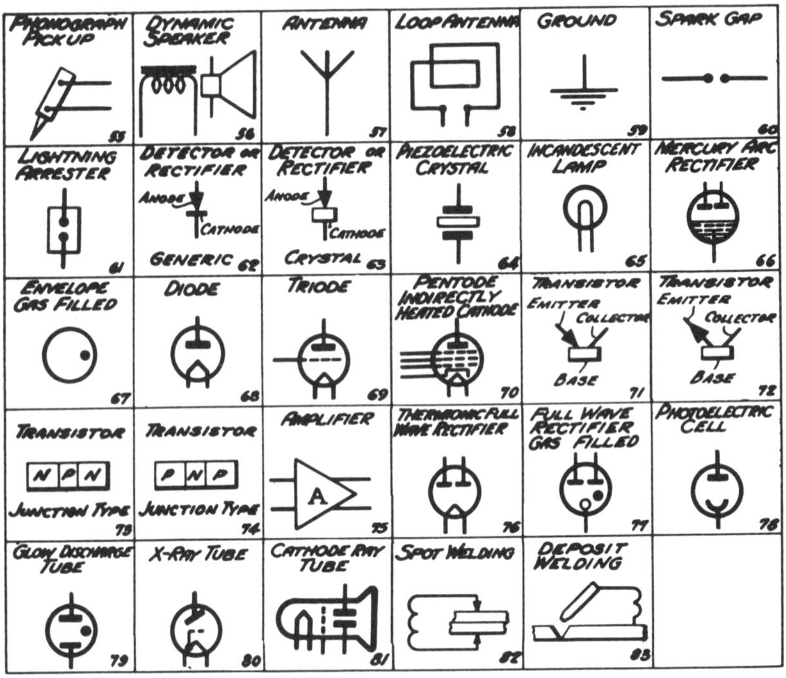

Mechanical Symbols

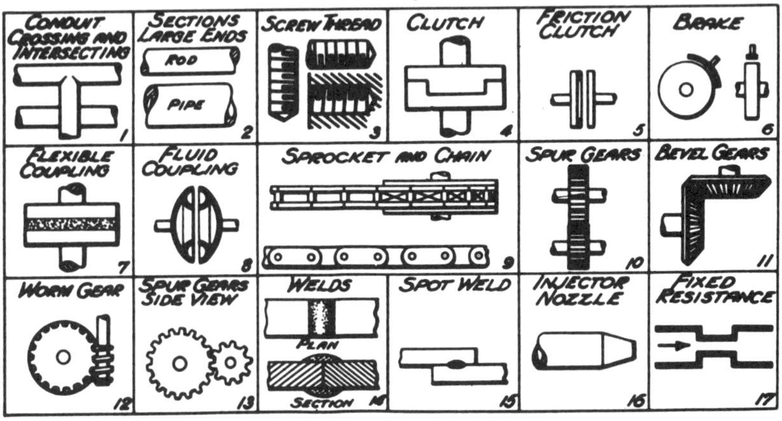

Figure 9-3 (*Continued*).
Approved graphical symbols for patent drawings.

Mechanical Symbols – continued

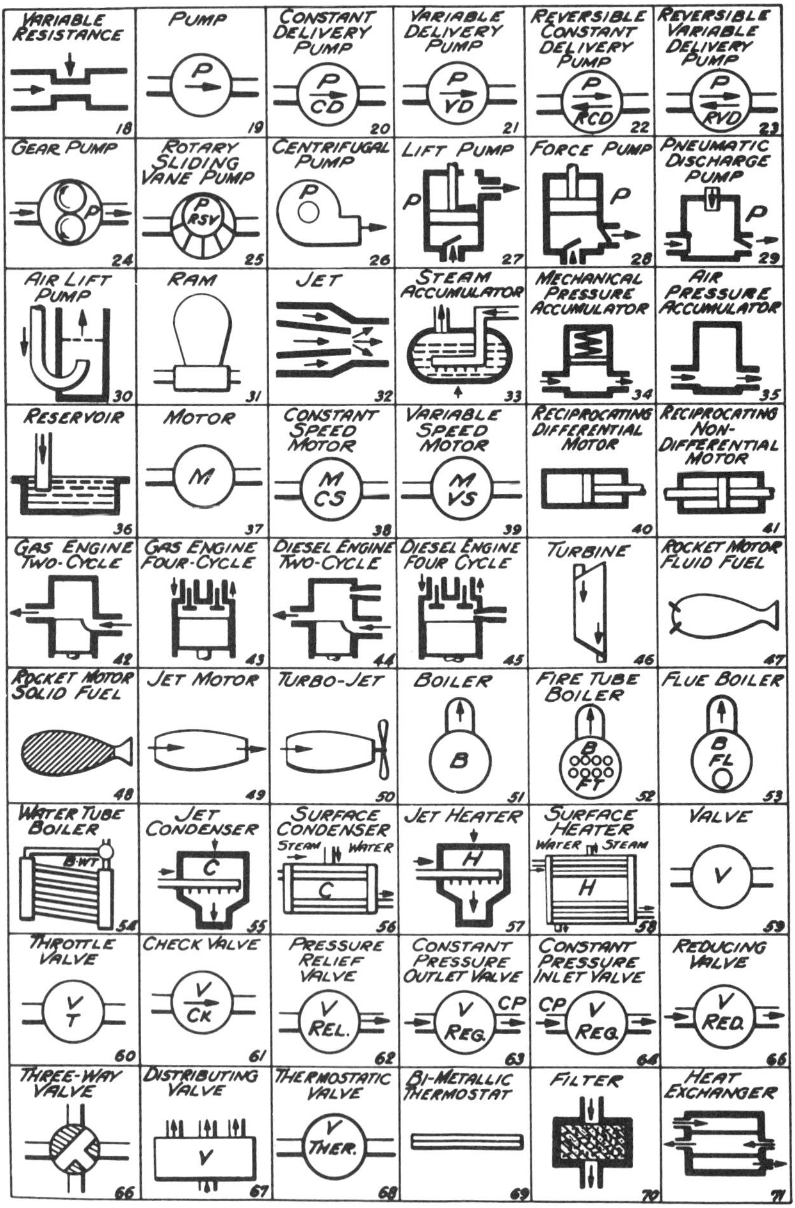

Figure 9-3 (*Continued*).
Approved graphical symbols for patent drawings.

PUBLICATIONS OF THE PATENT OFFICE

Patents

The specification and accompanying drawings of all patents are published on the day they are granted, and printed copies are sold to the public by the Patent Office. Over 3,000,000 patents have been issued.

Printed copies of any patent, identified by its patent number, may be purchased from the Patent Office at a cost of 50 cents, postage free, except design patents, which are 20 cents each.

Future patents classified in subclasses containing subject matter of interest may be obtained, as they issue, by prepayment of a deposit and a service charge. For the cost of such subscription service, a separate inquiry should be sent to the Patent Office.

Official Gazette of the United States Patent Office

The *Official Gazette* of the United States Patent Office is the official journal relating to patents and trademarks. It has been published weekly since January 1872 (replacing the old *Patent Office Reports*), and is now issued each Tuesday, simultaneously with the weekly issue of the patents. It contains a claim and a selected figure of the drawings of each patent granted on that day; decisions in patent and trademark cases rendered by the courts and the Patent Office; notices of patent and trademark suits; indexes of patents and patentees; list of patents available for license or sale; and much general information such as orders, notices, changes in rules, changes in classification, etc. The *Official Gazette* is sold, by annual subscription and in single copies, by the Superintendent of Documents, Washington, D.C.

Since July 1952, the illustrations and claims of the patents are arranged in order according to the Patent Office classification of subject matter. New patents relating to any particular field can be located readily. Street addresses of patentees have been published since May 24, 1960. A geographical index of the residences of inventors has been published since May 18, 1965. Copies of the *Official Gazette* may be found in public libraries of larger cities.

Index of Patents

This annual index to the *Official Gazette* contains an alphabetical index of the names of patentees and a list identifying the subject matter of the patents granted during the calendar year. Copies are sold by the Superintendent of Documents. At present it is issued in two volumes, one for patents and one for trademarks.

Decisions of the Commissioner of Patents

There is issued annually a volume republishing the decisions which have been published weekly in the *Official Gazette*. Copies are sold by the Superintendent of Documents.

Manual of Classification

This is a loose-leaf book containing a list of all the classes and subclasses of inventions in the Patent Office classification of patents, a subject matter index, and other information relating to classification. Substitute pages are issued from time to time. The manual and subscriptions for the substitute pages are sold by the Superintendent of Documents.

Classification Bulletins

The various changes and advances in classification made from time to time are collected and published in bulletins which give these changes and the definitions of new and revised classes and subclasses. These bulletins are sold by the Superintendent of Documents and the Patent Office.

Discovery of Registered Patent Attorneys and Agents Arranged by States and Cities

This list of registered patent attorneys and agents is arranged by states and cities where the registrants have offices. This publication is sold by the Superintendent of Documents.

Development and Use of Patent Classification Systems

This is a comprehensive guide setting forth the bases on which the U.S. Patent Classification System is presently organized, the techniques used in developing and administering the system, and instructions on how it may be used to obtain information. Copies are available from the Superintendent of Documents.

Manual of Patent Examining Procedure

This manual is written for examiners in the Patent Office and gives in great detail the procedures followed by the examiners. It is available to the public and is sold by the Superintendent of Documents.

Patent Laws

A compilation of the patent laws in force is issued, and revised editions are published from time to time. Copies may be purchased from the Superintendent of Documents.

Title 37—Patents, Trademarks and Copyrights

The contents of this volume represent all current regulations codified under the title of the CFR as of July 1, 1995.

Microfilm Lists of Patents

The reels containing the mechanical, electrical and chemical patent numbers in original and cross-referenced classification may be purchased from the Clearinghouse for Federal Scientific and Technical Information (CFSTI), 5283 Port Royal Road, Springfield, Va., 22151, to whom all communications respecting the same should be addressed.

FORMS

Figures 9-4 through 9-12 illustrate the manner of preparing parts of patent applications and other related papers.

Attorney's Docket No. ______________________ **PATENT**

IN THE UNITED STATES PATENT AND TRADEMARK OFFICE

Patent application

of ____

Inventor(s)

for ____

Title of invention

____

the specification of which is being transmitted herewith

OR

In re application of:
Serial No.: 0 /
Filed:
For:

Group No.:
Examiner:

Assistant Commissioner for Patents
Washington, D.C. 20231

INFORMATION DISCLOSURE STATEMENT

CERTIFICATION UNDER 37 CFR 1.8(a) and 1.10

I hereby certify that, on the date shown below, this correspondence is being:

MAILING

❑ deposited with the United States Postal Service in an envelope addressed to the Assistant Commissioner for Patents, Washington, D.C. 20231

37 CFR 1.8(a)

❑ with sufficient postage as first class mail.

37 CFR 1.10

❑ as "Express Mail Post Office to Addressee" Mailing Label No. ____________

TRANSMISSION

❑ transmitted by facsimile to the Patent and Trademark Office.

Date: ____________

____
Signature

____
(*type or print name of person certifying*)

(Information Disclosure Statement [6-1]—page 1 of 2)

Note: This form has been edited to show typical disclosure information required.

Figure 9-4. Information Disclosure Statement

NOTE: *"Each individual associated with the filing and prosecution of a patent application has a duty of candor and good faith in dealing with the Office, which includes a duty to disclose to the Office all information known to that individual to be material to patentability as defined in this section" 37 CFR 1.56(a).*

"Individuals associated with the filing or prosecution of a patent application within the meaning of this section are:

(1) each inventor named in the application;

(2) each attorney or agent who prepares or prosecutes the application; and

(3) every other person who is substantively involved in the preparation or prosecution of the application and who is associated with the inventor, with the assignee or with anyone to whom there is an obligation to assign the application." 37 CFR 1.56(c).

NOTE: *The "duty as described in § 1.56 will be met so long as the information in question was cited by the Office or submitted to the Office in the manner prescribed by §§ 1.97(b) - (d) and 1.98 before issuance of the patent." Notice of January 9, 1992, 1135, O.G. 13-25 at 17.*

WARNING: "No information disclosure statement may be filed in a provisional application." 37 CFR § 1.51(b).

List of Sections Forming Part of This Information Disclosure Statement

The following sections are being submitted for this information Disclosure Statement:
(check sections forming a part of this statement; discard unused sections and number pages consecutively)

1. ❑ Preliminary Statements
2. ❑ FORM PTO - 1449 (Modified)
3. ❑ Statement as to Information Not Found in Patents or Publications
4. ❑ Identification of Prior Application in Which Listed Information Was Already Cited and for Which No Copies Are Submitted or Need Be Submitted
5. ❑ Cumulative Patents or Publications
6. ❑ Copies of Listed Information Items Accompanying This Statement
7. ❑ Concise Explanation of Non-English Language Listed Information Items
   7A ❑ EPO Search Report
   7B. ❑ English Language Version of EPO Search Report
8. ❑ Translation(s) of Non-English Language Documents
9. ❑ Concise Explanation of English Language Listed Information Items (Optional)
10 ❑ Identification of Person(s) Making This Information Disclosure Statement

(complete the following, if appropriate)

Sections   , respectively, have been continued on ADDED PAGE(S).

NOTE: "Once the minimum requirements are met, the examiner has an obligation to consider the information." Notice of April 20, 1992 (1138 O.G. 37-41, 37).

(Information Disclosure Statement [6-1]—2 of 2)

Figure 9-4. (*Continued*)

Section 1. Preliminary statements

Applicants submit herewith patents, publications or other information, of which they are aware that they believe may be material to the examination of this application, and in respect of which there may be a duty to disclose.

The filing of this information disclosure statement shall not be construed as a representation that a search has been made (37 CFR 1.97(g)), an admission that the information cited is, or is considered to be, material to patentability, or that no other material information exists.

The filing of this information disclosure statement shall not be construed as an admission against interest in any manner. Notice of January 9, 1992, 1135 O.G. 13-25, at 25.

(Information Disclosure Statement — Section 1. Preliminary Statements [6-1] — page ____ of ____)

Figure 9-4. (*Continued*)

Approved for use through 9/30/98, OMB 0651-0031
Patent and Trademark Office: U.S. DEPARTMENT OF COMMERCE

Please type a plus sign (+) inside this box → ❑

1449A/PTO Rev. 10/95	U.S. Department of Commerce Patent and Trademark Office	*Complete If Known*	
LIST OF PRIOR ART CITED BY APPLICANT *(use as many sheets as necessary)*		Application Number	
		Filing Date	
		First Named Inventor	
		Group Art Unit	
		Examiner Name	
Sheet	of	Attorney Docket Number	

U.S. PATENT DOCUMENTS

Examiner Initials*	Cite No.[1]	U.S. Patent Document		Name of Patentee or Applicant of Cited Document	Date of Publication of Cited Document MM-DD-YYYY	Pages, Columns, Lines, Where Relevant Passages or Relevant Figures Appear
		Number	Kind Code *(if known)*			

FOREIGN PATENT DOCUMENTS

Examiner Initials*	Cite No.[1]	Foreign Patent Document			Name of Patentee or Applicant of Cited Document	Date of Publication of Cited Document MM-DD-YYYY	Pages, Columns, Lines, where Relevant Passages or Relevant Figures Appear	T[6]
		Office[3]	Number[4]	Kind Code[5] *(if known)*				

Examiner Signature		Date Considered	

(Information Disclosure Statement—Section 2. FORM 1449A/PTO [6-1]—page ____ of ____)

Figure 9-4. (*Continued*)

Approved for use through 9/30/98, OMB 0651-0031
Patent and Trademark Office: U.S. DEPARTMENT OF COMMERCE

Please type a plus sign (+) inside this box → ❑

1449B/PTO Rev. 10/95	U.S. Department of Commerce Patent and Trademark Office	*Complete If Known*	
LIST OF PRIOR ART CITED BY APPLICANT *(use as many sheets as necessary)*		Application Number	
		Filing Date	
		First Named Inventor	
		Group Art Unit	
		Examiner Name	
Sheet	of	Attorney Docket Number	

OTHER PRIOR ART — NON PATENT LITERATURE DOCUMENTS

Examiner Initials*	Cite No.[1]	Include name of the author (in CAPITAL LETTERS), title of the article (when appropriate), title of the item (book, magazine, journal, serial, symposium, catalog, etc.), date, page(s), volume-issue number(s), publisher, country, where published, source.)	T[2]

Examiner Signature		Date Considered	

(Information Disclosure Statement—Section 2. FORM 1449B/PTO [6-1]—page ____ of ____)

Figure 9-4. (*Continued*)

Section 4. Identification of Prior Application in Which Listed Information Was Already Cited and for Which No Copies Are Submitted or Need Be Submitted

Note: *"A copy of any patent, publication or other information listed in an information disclosure statement is not required to be provided if it was previously cited by or submitted to the Office in a prior application, provided that the prior application is properly identified in the statement and relied upon for an earlier filing date under 35 U.S.C. 120." 37 CFR 1.98(d).*

WARNING: *"This exception to the requirement for copies of information does not apply to information which was cited in an international application under the Patent Cooperation Treaty." Notice of April 20, 1992 (1138 O.G. 37-41, 38).*

WARNING: *Information which is cited or submitted to the Office in the parent application of a file wrapper continuing application under 37 CFR 1.62 will be part of the file before the examiner and need not be resubmitted in the continuing application. Likewise, the examiner will consider information cited or submitted to the Office in a parent application when examining a continuing application and thus this information need not be resubmitted unless applicant desires the information to be printed on the patent." Notice of April 20, 1992 (1138 O.G. 37-41, 37).*

WARNING: *While a copy of a non-English language item of information need not be submitted, if it was previously submitted to, or cited by, the Office in a prior application, provided it is properly identified in this statement and this application relies on that earlier filing date under 35 U.S.C. 120, nevertheless, the requirement in § 1.98(a)(3) for a concise explanation of non-English language information would not be satisfied by a statement that a reference was cited in the prosecution of a parent application. The concise explanation must explain the relevance, as presently understood by the person designated in § 1.56(c) most knowledgeable about the content of the information. Notice of January 9, 1992, 1135 O.G. 13-25, at 20.*

This application relies, under 35 U.S.C. 120, on the earlier filing date of prior application SN.: 0 / , filed on ____________ (date).

(complete the following, if applicable)

❑ This application also relies, under 35 U.S.C. 120, on the earlier filing date of prior application SN.: 0 / filed on ____________ (date).

The following references were submitted to, and/or cited by, the Office in the prior application(s) and, therefore, are not required to be provided in this application:

(Information Disclosure Statement—Section 4. Identification of Prior Application in Which Listed Information Was Already Submitted and for Which No Copies Are Submitted or Need Be Submitted [6-1]—page _____ of _____)

Figure 9-4. (*Continued*)

Section 5. Cumulative Patents or Publications

NOTE: *"When the disclosure of two or more patents or publications listed in an information disclosure statement are substantially cumulative, a copy of one of the patents or publications may be submitted without copies of the other patents or publications provided that a statement is made that these other patents or publications are cumulative." 37 CFR 1.98(c).*

NOTE: *"The examiner will then consider only the patent or publication of which a copy is submitted and will so indicate on the list or form PTO-1449 submitted, e.g., by crossing-out the listing of the cumulative information." Notice of April 20, 1992 (1138 O.G. 37-41, 38).*

STATEMENT

________________ is cumulative of the following patents or publications listed on Form PTO-1449 (Modified):

In accordance with 37 CFR 1.98(c), a copy of only ________________ is being submitted with this Information Disclosure Statement.

(Information Disclosure Statement— Section 5. Cumulative Patents or Publications [6-1] —page ____ of ____)

Figure 9-4. (*Continued*)

Section 6. Copies of Listed Information Items Accompanying This Statement

NOTE: *37 CFR 1.98(a)(2) requires that any information disclosure statement filed under § 1.97 shall include: "A legible copy of: (1) Each U.S. and foreign patent; (ii) Each publication or that portion which caused it to be listed; and (iii) All other information or that portion which caused it to be listed, except that no copy of a U.S. patent application need be included..."*

NOTE: *The wording in § 1.98(a)(2)(iii) makes it clear that the requirement to submit a copy of each item of information listed in an information disclosure statement does not apply to the citation of a U.S. patent application. Notice of January 9, 1992, 1135 O.G. 13-25, at 14.*

Legible copies of all items listed in Form PTO-1449 accompany this information statement.

(complete the following, if applicable)

❑ Exception(s) to above:

- ❑ Items in prior application, from which an earlier filing date is claimed for this application, as identified in Section 4.
- ❑ Cumulative patents or publications identified in Section 5.

(Information Disclosure Statement—Section 6. Copies of Listed Information Items Accompanying This Statement [6-1]—page ____ of ____)

Figure 9-4. (*Continued*)

Section 7. Concise Explanation of Non-English Language Listed Information Items

NOTE: *"A concise explanation of the relevance, as it is presently understood by the individual designated in § 1.56(c) most knowledgeable about the content of the information of each patent, publication, or other information listed that is not in the English language shall be included in any information disclosure statement filed under § 1.97. The concise explanation may be either separate from the specification or incorporated therein." 37 CFR 1.98(a)(3).*

NOTE: *"[T]he explanation required is limited to the relevance as understood by the individual designated in § 1.56(c) most knowledgeable about the content of the information at the time the information is submitted to the Office." Notice of January 9, 1992, 1135 O.G. 13-25 at 14.*

NOTE: *"Where the information listed is not in the English language, but was cited in a search report by a foreign patent officer, the requirement for a concise explanation of relevance is satisfied by submitting an English language version of the search report." Notice of January 9, 1992, 1135 O.G. 13-25, at 14.*

NOTE: *"The concise explanation requirement for non-English language information may be met by submission of an English language version of the search report indicating the degree of relevance found by the foreign office." Notice of January 9, 1992, 1135 O.G. 13-25, at 20.*

WARNING: "The requirement in § 1.98(a)(3) for a concise explanation of non-English language information would not be satisfied by a statement that a reference was cited in the prosecution of a parent application." Notice of January 9, 1992, 1135 O.G. 13-25 at 20 (emphasis added).

(Information Disclosure Statement—Section 7. Concise Explanation of Non-English Language Listed Information Items [6-1]—page ____ of ____)

Figure 9-4. (*Continued*)

Section 7A. Concise Explanation of Non-English Language Listed Information Items—EPO Search Report

The relevance with respect to the following citations listed on Form PTO-1449:

is submitted on the basis of the accompanying:

(check the appropriate item)

- ❑ EPO search report that is in the English language,
- ❑ EPO search report that is not in the English language and that is accompanied also by an English language version of the EPO search report,

that issued on the corresponding European patent application.

(Information Disclosure Statement—Section. 7A. Concise Explanation of Non-English Language Listed Information Items [6-1]—page ____ of ____)

Figure 9-4. (*Continued*)

Attorney's Docket No. ________________ *PATENT*

COMBINED DECLARATION AND POWER OF ATTORNEY

(ORIGINAL, DESIGN, NATIONAL STAGE OF PCT, SUPPLEMENTAL, DIVISIONAL, CONTINUATION OR C-I-P)

As a below named inventor, I hereby declare that:

TYPE OF DECLARATION

This declaration is of the following type:

(check one applicable item below)

- ❑ original.
- ❑ design.
- ❑ supplemental.

NOTE: *If the declaration is for an International Application being filed as a divisional, continuation or continuation-in-part application, do not check next item; check appropriate one of last three items.*

- ❑ national stage of PCT.

NOTE: *If one of the following 3 items apply, then complete and also attach ADDED PAGES FOR DIVISIONAL, CONTINUATION OR C-I-P.*

- ❑ divisional.
- ❑ continuation.
- ❑ continuation-in-part (C-I-P).

INVENTORSHIP IDENTIFICATION

WARNING: If the inventors are each not the inventors of all the claims, an explanation of the facts, including the ownership of all the claims at the time the last claimed invention was made, should be submitted.

My residence, post office address and citizenship are as stated below, next to my name. I believe that I am the original, first and sole inventor (*if only one name is listed below*) or an original, first and joint inventor (*if plural names are listed below*) of the subject matter that is claimed, and for which a patent is sought on the invention entitled:

TITLE OF INVENTION

____

____

(Declaration and Power of Attorney [1-1]–page 1 of 7)

Figure 9-5. Declaration and power of attorney.

SPECIFICATION IDENTIFICATION

the specification of which:

(complete (a), (b) or (c))

(a) ❑ is attached hereto.

NOTE: *"The following combinations of information supplied in an oath or declaration filed on the application filing date with a specification are acceptable as minimums for identifying a specification and compliance with any one of the items below will be accepted as complying with identification requirement of 37 CFR 1.63:*

"(1) name of inventor(s), and reference to an attached specification which is both attached to the oath or declaration at the time of execution and submitted with the oath or declaration on filing;

"(2) name of inventor(s), and attorney docket number which was on the specification as filed;

or

"(3) name of inventor(s), and title which was on the specification as filed."

Notice of July 13, 1995 (1177 O.G. 60).

(b) ❑ was filed on ________________, as ❑ Serial No. 0 / ______________

or ❑ ____

and was amended on ______________________ *(if applicable).*

NOTE: *Amendments filed after the original papers are deposited with the PTO that contain new matter are not accorded a filing date by being referred to in the declaration. Accordingly, the amendments involved are those filed with the application papers or, in the case of a supplemental declaration, are those amendments claiming matter not encompassed in the original statement of invention or claims. See 37 CFR 1.67.*

NOTE: *"The following combinations of information supplied in an oath or declaration filed after the filing date are acceptable as minimums for identifying a specification and compliance with any one of the items below will be accepted as complying with the identification requirement of 37 CFR 1.63:*

"(1) name of inventor(s), and application number (consisting of the series code and the serial number, e.g.,08/123,456);

"(2) name of inventor(s), serial number and filing date;

"(3) name of inventor(s) and attorney docket number which was on the specification as filed;

"(4) name of inventor(s) title which was on the specification as filed and filing date;

"(5) name of inventor(s), title which was on the specification as filed and reference to an attached specification which is both attached to the oath or declaration at the time of execution and submitted with the oath or declaration;

or

"(6) name of inventor(s), title which was on the specification as filed and accompanied by a cover letter accurately identifying the application for which it was intended by either the application number (consisting of the series code and the serial number; e.g., 08/123,456), or serial number and filing date. Absent any statement(s) to the contrary, it will be presumed that the application filed in the PTO is the application which the inventor(s) executed by signing the oath or declaration."

Notice of July 13, 1995 (1177 O.G. 60).

(c) ❑ was described and claimed in PCT International Application No. ____________, filed on ______________ and as amended under PCT Article 19 on __________ *(if any).*

(Declaration and Power of Attorney [1-1]–page 2 of 7)

Figure 9-5 (*Continued*)

ACKNOWLEDGEMENT OF REVIEW OF PAPERS AND DUTY OF CANDOR

I hereby state that I have reviewed and understand the contents of the above-identified specification, including the claims, as amended by any amendment referred to above.

I acknowledge the duty to disclose information, which is material to patentability as defined in 37, Code of Federal Regulations, § 1.56,

(also check the following items, if desired)

❑ and which is material to the examination of this application, namely, information where there is a substantial likelihood that a reasonable Examiner would consider it important in deciding whether to allow the application to issue as a patent, and

❑ in compliance with this duty, there is attached an information disclosure statement, in accordance with 37 CFR 1.98.

PRIORITY CLAIM (35 U.S.C. § 119(a)–(d))

I hereby claim foreign priority benefits under Title 35, United States Code, § 119(a)–(d) of any foreign application(s) for patent or inventor's certificate or of any PCT international application(s) designating at least one country other than the United States of America listed below and have also identified below any foreign application(s) for patent or inventor's certificate or any PCT international application(s) designating at least one country other than the United States of America filed by me on the same subject matter having a filing date before that of the application(s) of which priority is claimed.

(complete (d) or (e))

(d) ❑ no such applications have been filed.
(e) ❑ such applications have been filed as follows.

NOTE: *Where item(c) is entered above and the International Application which designated the U.S. itself claimed priority check item (e), enter the details below and make the priority claim.*

(Declaration and Power of Attorney [1-1]–page 3 of 7)

Figure 9-5 (*Continued*)

PRIOR FOREIGN/PCT APPLICATION(S) FILED WITHIN 12 MONTHS (6 MONTHS FOR DESIGN) PRIOR TO THIS APPLICATION AND ANY PRIORITY CLAIMS UNDER 35 U.S.C. § 119(a)-(d)

COUNTRY (OR INDICATE IF PCT)	APPLICATION NUMBER	DATE OF FILING (day, **month**, year)	PRIORITY CLAIMED UNDER 37 USC 119
			❑ YES NO ❑
			❑ YES NO ❑
			❑ YES NO ❑
			❑ YES NO ❑
			❑ YES NO ❑

CLAIM FOR BENEFIT OF PRIOR U.S. PROVISIONAL APPLICATION(S)
(34 U.S.C. § 119(e))

I hereby claim the benefit under Title 35, United States Code, § 119(e) of any United States provisional application(s) listed below:

PROVISIONAL APPLICATION NUMBER **FILING DATE**

_______/________________________ ____________

_______/________________________ ____________

_______/________________________ ____________

CLAIM FOR BENEFIT OF EARLIER US/PCT APPLICATION(S) UNDER 35 U.S.C. 120

❑ The claim for the benefit of any such applications are set forth in the attached ADDED PAGES TO COMBINED DECLARATION AND POWER OF ATTORNEY FOR DIVISIONAL, CONTINUATION OR CONTINUATION-IN-PART (C-I-P) APPLICATION.

(Declaration and Power of Attorney [1-1]—page 4 of 7)

Figure 9-5 (*Continued*)

ALL FOREIGN APPLICATION(S), *IF ANY,* FILED MORE THAN 12 MONTHS (6 MONTHS FOR DESIGN) PRIOR TO THIS U.S. APPLICATION

____

____

____

NOTE: *If the application filed more than 12 months from the filing date of this application is a PCT filing forming the basis for this application entering the United States as (1) the national stage, or (2) a continuation, divisional, or continuation-in-part, then also complete* ADDED PAGES TO COMBINED DECLARATION AND POWER OF ATTORNEY FOR DIVISIONAL, CONTINUATION OR C-I-P APPLICATION *for benefit of the prior U.S. or PCT application(s) under 35 U.S.C. § 120.*

POWER OF ATTORNEY

I hereby appoint the following attorney(s) and/or agent(s) to prosecute this application and transact all business in the Patent and Trademark Office connected therewith.

(list name and registration number)

(check the following item, if applicable)

❑ Attached, as part of this declaration and power of attorney, is the authorization of the above-named attorney(s) to accept and follow instructions from my representative(s).

SEND CORRESPONDENCE TO

DIRECT TELEPHONE CALLS TO:
(Name and telephone number)

DECLARATION

I hereby declare that all statements made herein of my own knowledge are true and that all statements made on information and belief are believed to be true; and further that these statements were made with the knowledge that willful false statements and the like so made are punishable by fine or imprisonment, or both, under Section 1001 of Title 18 of the United States Code, and that such willful false statements may jeopardize the validity of the application or any patent issued thereon.

(Declaration and Power of Attorney [1-1]–page 5 of 7)

Figure 9-5. (*Continued*)

SIGNATURE(S)

NOTE: *Carefully indicate the family (or last) name, as it should appear on the filing receipt and all other documents.*

Full name of sole or first inventor

______________	______________	______________
(GIVEN NAME)	*(MIDDLE INITIAL OR NAME)*	*FAMILY (OR LAST NAME)*

Inventor's Signature ______________________________

Date ______________ **Country of Citizenship** ______________

Residence ______________________________

Post Office Address ______________________________

____

Full name of second joint inventor, if any

______________	______________	______________
(GIVEN NAME)	*(MIDDLE INITIAL OR NAME)*	*FAMILY (OR LAST NAME)*

Inventor's Signature ______________________________

Date ______________ **Country of Citizenship** ______________

Residence ______________________________

Post Office Address ______________________________

____

Full name of third joint inventor, if any

______________	______________	______________
(GIVEN NAME)	*(MIDDLE INITIAL OR NAME)*	*FAMILY (OR LAST NAME)*

Inventor's Signature ______________________________

Date ______________ **Country of Citizenship** ______________

Residence ______________________________

Post Office Address ______________________________

____

(Declaration and Power of Attorney [1-1]—page 6 of 7)

Figure 9-5. (*Continued*)

(check proper box(es) for any of the following added page(s) that form a part of this declaration)

❑ **Signature** for fourth and subsequent joint inventors. *Number of pages added* ____________.

* * *

❑ **Signature** by administrator(trix), executor(trix) or legal representative for deceased or incapacitated inventor. *Number of pages added* ____________.

* * *

❑ **Signature** for inventor who refuses to sign or cannot be reached by person authorized under 37 CFR 1.47. *Number of pages added* ____________.

* * *

❑ Added page for **signature** by one joint inventor on behalf of deceased inventor(s) where legal representative cannot be appointed in time. (37 CFR 1.47)

* * *

❑ Added pages to combined declaration and power of attorney for divisional, continuation, or continuation-in-part (C-I-P) application.

❑ Number of pages added ____________

* * *

❑ Authorization of attorney(s) to accept and follow instructions from representative.

* * *

(If no further pages form a part of this Declaration, then end this Declaration with this page and check the following item)

❑ This declaration ends with this page.

(Declaration and Power of Attorney [1-1]–page 7 of 7)

Figure 9-5. (*Concluded*)

DECLARATION FOR PATENT APPLICATION	Docket Number (Optional)

As a below named inventor, I hereby declare that:
My residence, post office address and citizenship are as stated below next to my name.
I believe I am the original, first and sole inventor (if only one name is listed below) or an original, first and joint inventor (if plural names are listed below) of the subject matter which is claimed and for which a patent is sought on the invention entitled ______________________________, the specification of which is attached hereto unless the following box is checked:

❑ was filed on ________ as United States Application Number or PCT International Application Number ____________and was amended on _________________(if applicable).

I hereby state that I have reviewed and understand the contents of the above identified specification, including the claims, as amended by any amendment referred to above.
I acknowledge the duty to disclose information which is material to patentability as defined in 37 CFR § 1.56.
I hereby claim foreign priority benefits under 35 U.S.C. §119(a)-(d) or § 365(b) of any foreign application(s) for patent or inventor's certificate, or § 365(a) of any PCT international application which designated at least one country other than the United States, listed below and have also identified below, by checking the box, any foreign application for patent or inventor's certificate, or PCT International application having a filing date before that of the application on which priority is claimed.

Prior Foreign Application(s) Priority Not Claimed

______________ (Number)	______________ (Country)	______________ (Day/Month/Year Filed)	❑
______________ (Number)	______________ (Country)	______________ (Day/Month/Year Filed)	❑

I hereby claim the benefit under 35 U.S.C. § 119(e) of any United States provisional application(s) listed below.

______________ (Application Number)	______________ (Filing Date)
______________ (Application Number)	______________ (Filing Date)

I hereby claim the benefit under 35 U.S.C. § 120 of any United States application(s) or § 365(c) of any PCT International application designating the United States, listed below and, insofar as the subject matter of each of the claims of this application is not disclosed in the prior United States or PCT International application in the manner provided by the first paragraph of 35 U.S. C. § 112, I acknowledge the duty to disclose information which is material to patentability as defined in 37 CFR § 1.56 which became available between the filing date of the prior application and the national or PCT International filing date of this application.

______________ (Application Number)	______________ (Filing Date)	______________ (Status–Patented, Pending, Abandoned)
______________ (Application Number)	______________ (Filing Date)	______________ (Status–Patented,Pending, Abandoned)

I hereby appoint the following attorney(s) and/or agent(s) to prosecute this application and to transact all business in the Patent and Trademark Office connected therewith:

____

Address all telephone calls to ________________ at telephone number ________________
Address all correspondence to ______________________________

I hereby declare that all statements made herein of my own knowledge are true and that all statements made on information and belief are believed to be true; and further that these statements were made with the knowledge that willful false statements and the like so made are punishable by fine or imprisonment, or both, under Section 1001 of Title 18 of the United States Code and that such willful false statements may jeopardize the validity of the application or any patent issued thereon.

Full name of sole or first inventor (given name, family name)________________
Inventors Signature ________________Date ________________
Residence ________________ Citizenship ________________
Post Office Address ________________
Full name of second joint inventor, if any (given name, family name)________________
Second inventor;s signature ________________ Date________________
Residence ________________ Citizenship ________________
Post Office Address ________________

❑ Additional inventors are being named on separately numbered sheets attached hereto.

Declaration for Patent Application (PTO/SB/01) [1-1.1]–page 1 of 2)

Figure 9-6.

Type a plus sign (+) inside this box → ❑

0010/PTO Rev. 6/95 — U.S. Department of Commerce Patent and Trademark Office	Attorney Docket Number	
DECLARATION FOR UTILITY OR DESIGN PATENT APPLICATION	First Named Inventor	
	COMPLETE IF KNOWN	
	Application Number	
❑ Declaration Submitted with Initial Filing OR ❑ Declaration Submitted after Initial Filing	Filing Date	
	Group Art Unit	
	Examiner Name	

As a below named Inventor, I hereby declare that:

My residence, post office address, and citizenship are as stated below next to my name.

I believe I am the original, first and sole inventor (if only one name is listed below) or an original, first and joint inventor (if plural names are listed below) of the subject matter which is claimed and for which a patent is sought on the invention entitled:

(Title of the Invention)

The specification of which

❑ is attached hereto

OR

❑ was filed on (MM/DD/YYYY) ______ as United States Application Number or PCT International Application Number ______ and was amended on (MM/DD/YYYY) ______ (if applicable).

I hereby state that I have reviewed and understand the contents of the above identified specification, including the claims, as amended by any amendment specifically referred to above.

I acknowledge the duty to disclose information which is material to patentability as defined in Title 37 Code of Federal Regulations, § 1.56.

I hereby claim foreign priority benefits under Title 35, United States Code §119 (a)-(d) or § 365(b) of any foreign application(s) for patent or inventor's certificate, or § 365(a) of any PCT international application which designated at least one country other than the United States of America, listed below and have also identified below, by checking the box, any foreign application for patent or inventor's certificate, or of any PCT international application having a filing date before that of the application on which priority is claimed.

Prior Foreign Application Number(s)	Country	Foreign Filing Date (MM/DD/YYYY)	Priority Not Claimed	Certified Copy Attached? YES	NO
			❑	❑	❑
			❑	❑	❑
			❑	❑	❑
			❑	❑	❑
			❑	❑	❑
			❑	❑	❑

❑ Additional foreign application numbers are listed on a supplemental priority sheet attached hereto:

I hereby claim the benefit under Title 35, United States Code § 119(e) of any United States provisional application(s) listed below.

Application Number(s)	Filing Date (MM/DD/YYYY)	❑ Additional provisional application numbers are listed on a supplemental priority sheet attached hereto.

(Declaration for Utility or Design Patent Application (PTO/SB/01) [1-10.1]–page 1 of __)

Figure 9-6a.

Type a plus sign (+) inside this box → ❑

DECLARATION	Page 2

I hereby claim the benefit under Title 35, United States Code § 120 of any United States application(s), or §365(c) of any PCT international application designating the United States of America, listed below and, insofar as the subject matter of each of the claims of this application is not disclosed in the prior United States or PCT International application in the manner provided by the first paragraph of Title 35, United States Code § 112. I acknowledge the duty to disclose information which is material to patentability as defined in Title 37, Code of Federal Regulations § 1.56 which became available between the filing date of the prior application and the national or PCT international filing date of this application.

U.S. Parent Application Number	PCT Parent Number	Parent Filing Date (MM/DD/YYYY)	Parent Patent Number *(if applicable)*

❑ Additional U.S. or PCT international application numbers are listed on a supplemental priority sheet attached hereto

As a named inventor, I hereby appoint the following attorney(s) and/or agent(s) to prosecute this application and to transact all business in the Patent and Trademark Office connected therewith:

❑ Firm Name ______ Customer Number or label ______
OR
❑ List attorney(s) and or agent(s) name and registration number below:

Name	Registration Number	Name	Registration Number

❑ Additional attorney(s) and/or agent(s) named on a supplemental sheet attached hereto.

Please direct all cooerespondence to: ❑ Customer Number or label ______ OR ❑ Fill in correspondence address below

Name					
Address					
Address					
City		State		ZIP	
Country		Telephone		Fax	

I hereby declare that all statements made herein of my own knowledge are true and that all statements made on information and belief are believed to be true; and further that these statements were made with the knowledge that willful false statements and the like so made are punishable by fine or imprisonment, or both, under Section 1001 of Title 18 of the United States Code and that such willful false statements may jeopardize the validity of the application or any patent issued thereon.

Name of Sole or First Inventor: ❑ A petition has been filed for this unsigned inventor

Given Name		Middle Initial		Family Name		Suffix e.g. Jr.	
Inventor's Signature						Date	
Residence: City		State		Country		Citizenship	
Post Office Address							
Post Office Address							
City		State		Zip		Country	

City		State		Zip		Country		Applicant Authority	

❑ Additional inventors are being named on supplemental sheet(s) attached hereto

(Declaration for Utility or Design Patent Application (PTO/SB/01) [1-10.1]–page 2 of __)

Figure 9-6a. (*Continued*)

Type a plus sign (+) inside this box →❑

DECLARATION	ADDITIONAL INVENTOR(S) Supplemental Sheet

Name of Additional Joint Inventor, if any:				❑ A petition has been filed for this unsigned inventor						
Given Name		Middle Initial		Family Name			Suffix e.g., Jr.			
Inventor's Signature					Date					
Residence: City		State		Country		Citizenship				
Post Office Address										
Post Office Address										
City		State		Zip		Country		Applicant Authority		
Name of Additional Joint Inventor, if any:				❑ A petition has been filed for this unsigned inventor						
Given Name		Middle Initial		Family Name			Suffix e.g., Jr.			
Inventor's Signature					Date					
Residence: City		State		Country		Citizenship				
Post office Address										
Post Office Address										
City		State		Zip		Country		Applicant Authority		

(Declaration for Utility or Design Patent Application (PTO/SB/01) [1-10.1]—page __ of __)

Figure 9-6a. (*Continued*)

Type a plus sign (+) inside this box →❑

DECLARATION	PRIORITY DATA (Supplemental Sheet)

Additional foreign applications:

Prior Foreign Application Number(s)	Country	Foreign Filing Date (MM/DD/YYYY)	Priority Not Claimed	Certified Copy Attached? Yes	No
			❑	❑	❑
			❑	❑	❑
			❑	❑	❑
			❑	❑	❑
			❑	❑	❑
			❑	❑	❑
			❑	❑	❑
			❑	❑	❑
			❑	❑	❑
			❑	❑	❑
			❑	❑	❑
			❑	❑	❑

Additional provisional applications:

Application Number	Filing Date (MM/DD/YYYY)

Additional U.S. applications:

U.S. Parent Application Number	PCT Parent Number	Parent Filing Date (MM/DD/YYYY)	Parent Patent Number *(if applicable)*

(Declaration for Utility or Design Patent Application (PTO/SB/01)[1-10.1]—page_ of _)

Figure 9-6a. (*Continued*)

Type a plus sign (+) inside this box →❑

DECLARATION	ATTORNEY and/or AGENT INFORMATION (Supplemental Sheet)		
Name	Registration Number	Name	Registration Number

Declaration for Utility or Design Patent Application (PTO/SB/01 [1.10.1]—page_ of _)

Figure 9-6a. (*Continued*)

Attorney's Docket No. ______________________ *PATENT*

COMBINED DECLARATION AND POWER OF ATTORNEY (PLANT PATENT APPLICATION)

As a below named inventor, I hereby declare that:

My residence, post office address and citizenship are as stated below next to my name, I believe I am the original, first and sole inventor (*if only one name is listed below*) or an original, first and joint inventor (*if plural names are listed below*) of the subject matter that is claimed and for which a patent is sought of the new and distinct variety of

____

____

the specification of which:

(a) ❑ is attached hereto.
(b) ❑ was filed on ____________ as Application Serial No.
0 / ____________ and was amended____________
(if applicable).

AVERMENT UNDER 37 CFR 1.162

I have asexually reproduced the new and distinct variety.
❑ The new and distinct variety was found in a cultivated area.
NOTE: This item must be checked where the plant is a newly found plant.

ACKNOWLEDGEMENT OF REVIEW OF PAPERS AND DUTY OF CANDOR

I hereby state that I have reviewed and understand the contents of the above identified specification, including the claims, as amended by any amendment referred to above.

I acknowledge the duty to disclose information that is material to the examination of this application namely, information where there is a substantial likelihood that a reasonable examiner would consider it important in deciding whether to allow the application to issue as a patent.

❑ In compliance with this duty, there is attached an information disclosure statement in accordance with 37 CFR 1.98.

PRIORITY CLAIM

I hereby claim foreign priority benefits under Title 35, United States Code, § 119 of any foreign application(s) for patent or inventor's certificate listed below and have also identified below any foreign application for patent or inventor's certificate having a filing date before that of the application on which priority is claimed.

(complete (c) or (d))

(c) ❑ No such applications have been filed.
(d) ❑ Such applications have been filed as follows:

(Plant Patent Application and Power of Attorney [1-11]–page 1 of 4)

Figure 9-7.

Earliest Foreign Application(s), if any, Filed Within 12 Months Prior to Said Application

Country	Application No.	Date of Filing (day, **month**, year)	Date of issue (day, **month**, year)	Priority Claimed

All Foreign Application(s), if any, Filed More Than 12 Months Prior to Said Application

CONTINUATION-IN-PART

(complete this part only if this is a continuation-in-part application)

I hereby claim the benefit under Title 35, United States Code, § 120 of any United States application(s) listed below and, insofar as the subject matter of each of the claims of this application is not disclosed in the prior United States application in the manner provided by the first paragraph of Title 35, United States Code, § 112, I acknowledge the duty to disclose material information, namely, information where there is a substantial likelihood that a reasonable examiner would consider it important in deciding whether to allow the application to issue as a patent, which occurred between the filing date of the prior application and the national or PCT international filing date of this application:

0 /__________________________________
Application Serial No. Filing Date Status (patented, pending abandoned)

0 /__________________________________
Application Serial No. Filing Date Status (patented, pending abandoned)

POWER OF ATTORNEY

As a named inventor, I hereby appoint the following attorney(s) and/or agent(s) to prosecute this application and transact all business in the Patent and Trademark Office connected therewith.

(list name and registration number)

(Plant Patent Application and Power of Attorney [1-11]–page 2 of 4)

Figure 9-7. (*Continued*)

Type a plus sign (+) inside this box → ❑

Patent and Trademark Office: U.S. Department of Commerce

0010/PTO Rev. 6/95 — U.S. Department of Commerce Patent and Trademark Office	Attorney Docket Number	
	First Named Inventor	
PLANT PATENT APPLICATION (35 USC §161) DECLARATION	*COMPLETE IF KNOWN*	
	Application Number	
	Filing Date	
❑ Declaration Submitted with Initial Filing OR ❑ Declaration Submitted after Initial Filing	Group Art Unit	
	Examiner Name	

As a below named Inventor, I hereby declare that:

My residence, post office address, and citizenship are as stated below next to my name.

I believe I am the original, first and sole inventor (if only one name is listed below) or an original, first and joint inventor (if plural names are listed below) of the new and distinct variety of:

plant named:

which is claimed and for which a plant patent is sought,

the specification of which ❑ is attached hereto ❑ was filed on (MM/DD/YYYY) ______ as United States Application Number ______ and was amended on (MM/DD/YYYY) ______ (if applicable).

I hereby state that I have reviewed and understand the contents of the above identified specification, including the claims, as amended by any amendment specifically referred to above.
I have asexually reproduced the plant to which this application applies.

❑ Said plant was found in a cultivated area *(check this box for newly found plant only)*

I acknowledge the duty to disclose information which is material to patentability as defined in Title 37 Code of Federal Regulations, § 1.56.

I hereby claim foreign priority benefits under Title 35, United States Code §119 (a)-(d) or § 365(b) of any foreign application(s) for patent or inventor's certificate, or § 365(a) of any PCT international application which designated at least one country other than the United States of America, listed below and have also identified below, by checking the box, any foreign application for patent or inventor's certificate, or of any PCT international application having a filing date before that of the application on which priority is claimed.

Prior Foreign Application Number(s)	Country	Foreign Filing Date (MM/DD/YYYY)	Priority Not Claimed	Certified Copy Attached? YES	NO
			❑	❑	❑
			❑	❑	❑
			❑	❑	❑
			❑	❑	❑
			❑	❑	❑

❑ Additional foreign application numbers are listed on a supplemental sheet attached hereto:

I hereby claim the benefit under Title 35, United States Code § 119(e) of any United States provisional application(s) listed below.

Application Number(s)	Filing Date (MM/DD/YYYY)	❑ Additional provisional application numbers are listed on a supplemental sheet(s) attached hereto.

(Declaration for Plant Patent Application (PTO/SB/03) [1-11.1]–page 1 of __)

Figure 9-8.

Type a plus sign (+) inside this box → ❑

DECLARATION	Page 2

I hereby claim the benefit under Title 35, United States Code § 120 of any United States application(s), or §365(c) of any PCT international application designating the United States of America, listed below, and, insofar as the subject matter of the claim of this application is not disclosed in the prior United States or PCT international application in the manner provided by the first paragraph of Title 35, United States Code § 112, I acknowledge the duty to disclose information which is material to patentability as defined in Title 37, Code of Federal Regulations § 1.56 which became available between the filing date of the prior application and the national or PCR international filing date of this application.

U.S. Parent Application Number	PCT Parent Number	Parent Filing Date (MM/DD/YYYY)	Parent Patent Number (*if applicable*)

❑ Additional U.S. or PCT international application numbers are listed on a supplemental priority sheet attached hereto

As a named inventor, I hereby appoint the following attorney(s) and/or agent(s) to prosecute this application and to transact all business in the Patent and Trademark Office connected therewith:

❑ Firm Name [ ] Customer Number or label [ ]
OR
❑ List attorney(s) and/or agent(s) name and registration number below:

Name	Registration Number	Name	Registration Number

❑ Additional attorney(s) and/or agent(s) named on a supplemental sheet attached hereto.

Please direct all correspondence to: ❑ Customer Number or label [ ] OR ❑ Fill in correspondence address below

Name					
Address					
Address					
City		State		ZIP	
Country		Telephone		Fax	

I hereby declare that all statements made herein of my own knowledge are true and that all statements made on information and belief are believed to be true; and further that these statements were made with the knowledge that willful false statements and the like so made are punishable by fine or imprisonment, or both, under Section 1001 of Title 18 of the United States Code and that such willful false statements may jeopardize the validity of the application of any patent issued thereon.

Name of Sole or First Inventor: ❑ A petition has been filed for this unsigned inventor

Given Name		Middle Initial		Family Name		Suffix e.g. Jr.	
Inventor's Signature				Date			
Residence: City		State		Country		Citizenship	
Post Office Address							
Post Office Address							
City		State		Zip		Country	
Applicant Authority							

❑ Additional inventors are being named on supplemental sheet(s) attached hereto

(Declaration for Plant Patent Application (PTO/SB/03) [1-11.1]–page 2 of __)

Figure 9-8. (*Continued*)

DECLARATION	ADDITIONAL INVENTOR(S) Supplemental Sheet

Name of Additional Joint Inventor, if any:	❑ A petition has been filed for this unsigned inventor								
Given Name		Middle Initial		Family Name		Suffix e.g., Jr.			
Inventor's Signature		Date							
Residence: City		State		Country		Citizenship			
Post Office Address									
Post Office Address									
City		State		Zip		Country		Applicant Authority	
Name of Additional Joint Inventor, if any:	❑ A petition has been filed for this unsigned inventor								
Given Name		Middle Initial		Family Name		Suffix e.g., Jr.			
Inventor's Signature		Date							
Residence: City		State		Country		Citizenship			
Post Office Address									
Post Office Address									
City		State		Zip		Country		Applicant Authority	

(Declaration for Plant Patent Application (PTO/SB/03) [1-11.1]—page __ of __)

Figure 9-8. (*Continued*)

Type a plus sign (+) inside this box →❑

DECLARATION	PRIORITY DATA (Supplemental Sheet)

Additional foreign applications:

Prior Foreign Application Number(s)	Country	Foreign Filing Date (MM/DD/YYYY)	Priority Not Claimed	Certified Copy Attached? Yes	No
			❑	❑	❑
			❑	❑	❑
			❑	❑	❑
			❑	❑	❑
			❑	❑	❑
			❑	❑	❑
			❑	❑	❑
			❑	❑	❑
			❑	❑	❑
			❑	❑	❑
			❑	❑	❑
			❑	❑	❑

Additional provisional applications:

Application Number	Filing Date (MM/DD/YYYY)

Additional U.S. applications:

U.S. Parent Application Number	PCT Parent Number	Parent Filing Date (MM/DD/YYYY)	Parent Patent Number *(if applicable)*

(Declaration for Plant Patent Application (PTO/SB/03)[1-11.1]—page_ of _)

Figure 9-8. (*Continued*)

Type a plus sign (+) inside this box →❑

DECLARATION		ATTORNEY and/or AGENT INFORMATION (Supplemental Sheet)	
Name	Registration Number	Name	Registration Number

Declaration for Plant Patent Application (PTO/SB/03, [1-11.1]—page_ of _)

Figure 9-8. (*Continued*)

(check the following item, if applicable)

❑ Attached as part of this declaration and power of attorney is the authorization of the above-named attorney(s) to accept and follow instructions from my representatives.

Send Correspondence To

Direct Telephone Calls to:
(name and telephone number)

I hereby declare that all statements made herein of my own knowledge are true and that all statements made on information and belief are believed to be true; and further that these statements were made with the knowledge that willful false statements and the like so made are punishable by fine or imprisonment, or both, under Section 1001 of Title 18 of the United States Code, and that such willful false statements may jeopardize the validity of the application or any patent issued thereon.

Full name of **sole or first** inventor ____________________

Inventor's signature ____________________

Date ____________ Country of Citizenship ____________

Residence ____________________

Post Office Address ____________________

Full name of **second joint inventor**, if any ____________________

Inventor's signature ____________________

Date ____________ Country of Citizenship ____________

Residence ____________________

Post Office Address ____________________

(CHECK PROPER BOX(ES) FOR ANY ADDED PAGE(S) FORMING A PART OF THIS DECLARATION)

❑ Signature for third and subsequent joint inventors. Number of pages added ____________.

(Plant Patent Application and Power of Attorney [1-11]—page 3 of 4)

Figure 9-8. (*Continued*)

- ❑ Signature by administrator(trix), executor(trix) or legal representative for deceased or incapacitated inventor. Number of pages added ____________.

- ❑ Signature for inventor who refuses to sign or cannot be reached by person authorized under 37 CFR 1.47. Number of pages added ______________.

- ❑ Authorization of attorney(s) to accept and follow instructions from representative.

(if no further pages form a part of this Declaration, then end this Declaration, with this page and check the following item)

❑ **This declaration ends with this page.**

(Plant Patent Application and Power of Attorney [1-11]—page 4 of 4

Figure 9-8. (*Continued*)

Attorney's Docket No. ____________________ ***PATENT***

❑ Applicant	❑ Patentee ____________________
❑ Application No.	❑ Patent No. ____________________
❑ Filed on	❑ Issued on ____________________

Title: ____

VERIFIED STATEMENT CLAIMING SMALL ENTITY STATUS
(37 CFR 1.9(f) and 1.27(b))—INDEPENDENT INVENTOR

As a below named inventor, I hereby declare that I qualify as an independent inventor, as defined in 37 CFR 1.9(c), for purposes of paying reduced fees to the United States Patent and Trademark Office under Sections 41(a) and (b) of Title 35, United States Code, to the Patent and Trademark Office, with regard to the invention described in

❑ the specification filed herewith, with title as listed above.
❑ the application identified above.
❑ the patent identified above.

I have not assigned, granted, conveyed or licensed, and am under no obligation under contract or law to assign, grant, convey or license, any rights in the invention to any person who would not qualify as an independent inventor under 37 CFR 1.9(c), if that person had made the invention, or to any concern that would not qualify as a small business concern under 37 CFR 1.9(d), or a nonprofit organization under 37 CFR 1.9(e).

Each person, concern or organization to which I have assigned, granted, conveyed, or licensed or am under an obligation under contract or law to assign, grant, convey, or license any rights in the invention is listed below:

❑ No such person, concern, or organization exists.
❑ Each such person, concern or organization is listed below.*

**NOTE: Separate verified statements are required from each named person, concern or organization having rights to the invention averring to their status as small entities. (37 CFR 1.27)*

FULL NAME ____
ADDRESS ____
____

❑ INDIVIDUAL ❑ SMALL BUSINESS CONCERN ❑ NONPROFIT ORGANIZATION

FULL NAME ____
ADDRESS ____
____

❑ INDIVIDUAL ❑ SMALL BUSINESS CONCERN ❑ NONPROFIT ORGANIZATION

Small Entity—Independent Inventor [7-1]—page 1 of 2)

Figure 9-9.

I acknowledge the duty to file, in this application or patent, notification of any change in status resulting in loss of entitlement to small entity status prior to paying, or at the time of paying, the earliest of the issue fee or any maintenance fee due after the date on which status as a small entity is no longer appropriate. (37 CFR 1.28(b))

I hereby declare that all statements made herein of my own knowledge are true and that all statements made on information and belief are believed to be true; and further, that these statements were made with the knowledge that willful false statements and the like so made are punishable by fine or imprisonment, or both, under Section 1001 of Title 18 of the United States Code, and that such willful false statements may jeopardize the validity of the application, any patent issuing thereon, or any patent to which this verified statement is directed.

____

Name of Inventor

______________________________ Date __________________

Signature of Inventor

____

Name of Inventor

______________________________ Date __________________

Signature of Inventor

____

Name of Inventor

______________________________ Date __________________

Signature of Inventor

(Small Entity—Independent Inventor [7-1]—page 2 of 2)

Figure 9-9. (*Continued*)

Attorney's Docket No. ______________________ *PATENT*

IN THE UNITED STATES PATENT AND TRADEMARK OFFICE

❑ **In re application of*:**
Serial No.: / Group No:
Filed: Examiner:
For*:
❑ **Patent No.:** Issued:

**NOTE: Insert name(s) of all inventor(s) and title also for patent.*

Assistant Commissioner for Patents
Washington, D.C. 20231

POWER OF ATTORNEY BY INVENTOR(S)
(REVOCATION OF PRIOR POWERS)

As a named inventor for the above identified

❑ application,
❑ patent,

REVOCATION OF PRIOR POWERS OF ATTORNEY

I hereby revoke all powers of attorney previously given and

NEW POWER OF ATTORNEY

I hereby appoint the following attorney(s) and/or agent(s) to prosecute and transact all business in the Patent and Trademark Office connected therewith.

(list name(s) and registration number(s))

(check the following item, if applicable)

❑ Attached as part of this power of attorney is the authorization of the above-named attorney(s) to accept and follow instructions from my representative(s).

SEND CORRESPONDENCE TO: **DIRECT TELEPHONE CALLS TO:**

(Power of Attorney by Inventor(s) [12-1]—page 1 of 2)

Figure 9-10.

	______________________
	(Type or print inventor's name)
Date: ______________	
	______________________
	Inventor's signature
	______________________
	Post Office Address
	______________________

	______________________
	(Type or print inventor's name)
Date: ______________	
	______________________
	Inventor's signature
	______________________
	Post Office Address
	______________________

	______________________
	(Type or print inventor's name)
Date: ______________	
	______________________
	Inventor's signature
	______________________
	Post Office Address
	______________________

NOTE: A power of attorney may be revoked at any stage in the prosecution of a case.

- ❑ Added page(s) forming a part of this power of attorney
- ❑ Added page(s) for signature(s) by additional inventors
- ❑ Added page—Authorization of attorney(s) to accept and follow instructions from representative

(supply similar information and signature for fourth and subsequent inventors)

(Power of Attorney by Inventor(s) [12-1]—page 2 of 2)

Figure 9-10. (*Continued*)

ADDED PAGE TO POWER OF ATTORNEY BY INVENTOR(S) FOR AUTHORIZATION OF ATTORNEY(S) TO ACCEPT AND FOLLOW INSTRUCTIONS FROM REPRESENTATIVE

The undersigned to this power of attorney hereby authorize(s) the U.S. attorney(s) named herein to accept and follow instructions from

____

Name(s) of authorized representative(s)

____

Address

____

as to any actions to be taken in the Patent and Trademark Office regarding this application without direct communication between the United States attorney(s) and the undersigned. In the event of a change in the person(s) from whom instructions may be taken, the U.S. attorney(s) will be so notified by the undersigned.

(Added page to Power of Attorney by Inventor(s) to Accept and Follow Instructions From Representative—[12-1])

Figure 9-10. (*Continued*)

Approved for use through 10/31/95. OMB 0651-0027
Patent and Trademark Office; U.S. DEPARTMENT OF COMMERCE

ASSIGNMENT OF APPLICATION	Docket Number (Optional)

Whereas, I, ____________________ of ________________, hereafter referred to as applicant, have invented certain new and useful improvements in ______________________________
____

❑ for which an application for a United States Patent was filed on __________________.
Application Number _______/________________.

❑ for which an application for a United States Patent was executed on __________, and
Whereas, ________________________ of ______________________ herein referred to as "assignee" whose post office address is ____,
is desirous of acquiring the entire right, title and interest in the same;
Now, therefore, in consideration of the sum of __________ dollars ($__________), the receipt whereof is acknowledged, and other good and valuable consideration, I, the applicant, by these presents do sell, assign and transfer unto said assignee the full and exclusive right to the said invention in the United States and the entire right, title and interest in and to any and all Patents which may be granted therefor in the United States, I hereby authorized and request the Commissioner of Patents and Trademarks to issue said United States Patent to said assignee, of the entire right, title, and interest in and to the same, for his sole use and behoof; and for the use and behoof of his legal representatives, to the full end of the term for which said Patent may be granted, as fully and entirely as the same would have been held by me had this assignment and sale not been made.
Executed this ____________________________ day of ____________________, 19 __________,
at ____.

____
(Signature)

State of _____________)
County of ___________) SS:
Before me personally appeared said ________________________________
and acknowledged the foregoing instrument to be his free act and deed
this ____________ day of _____________, 19 ______.

Seal ________________________________
(Notary Public)

(Assignment of Application (PTO/SB/15) [16-3.3]—page 1 of 1)

Figure 9-11.

Approved for use through 10/31/95. OMB 0651-0027
Patent and Trademark Office; U.S. DEPARTMENT OF COMMERCE

ASSIGNMENT OF PATENT	Docket Number (optional)

Whereas, I, ________________ of ______________, hereinafter referred to as patentee, did obtain a United States Patent for an improvement in ______________________

____

No. ________________, dated ________________; and whereas, I am now the sole owner of said patent, and,

Whereas, ______________________________________

of ____

hereinafter referred to as "assignee" whose post office address is ____________________

____

City of ________________________, and State of ____________________

is desirous of acquiring the entire right, title and interest in the same;

Now, therefore, in consideration of the sum of ________ dollars ($__________), the receipt whereof is acknowledged, and other good and valuable consideration, I, the patentee, by these presents do sell, assign and transfer unto said assignee the entire right, title and interest in and to the said Patent aforesaid; the same to be held and enjoyed by the said assignee for his own use and behoof, and for his legal representatives and assigns to the full end of the term for which said Patent is granted, as fully and entirely as the same would have been held by me had this assignment and sale not been made.

Executed this ____________________ day of ______________, 19 ________

at ____

State of ____________)
County of ___________) SS:

Before me personally appeared said ________________________

and acknowledged the foregoing instrument to be his free act and deed

this __________ day of __________, 19 ______.

Seal

____

(Notary Public)

(Assignment of Patent (PTO/SB/41) [16-1]—page 1 of 1)

Figure 9-12.

INDEX